MATTER

Open and shut mould
for making apothecaries'
phials (19th century)

Metals produced by
electrolysis (mid 19th century)

Crookes' thallium compounds
and notebook (1860s)

Measuring
cylinder
(19th century)

Model of adenovirus
(20th century)

Ammonium
dichromate crystals

Bunsen burner with
fan (19th century)

Graphite

Roman coins
(2nd century)

Boxes of chemicals
(19th century)

EYEWITNESS 👁 GUIDES

MATTER

Written by
CHRISTOPHER COOPER

Tyndall's boiling
point apparatus
(1880s)

Tongs and calcium
oxide powder

Heating limestone
(calcium carbonate)

DORLING KINDERSLEY
London • New York • Moscow • Sydney
in association with
THE SCIENCE MUSEUM, LONDON

Boxes of chemicals (19th century)

DK

A DORLING KINDERSLEY BOOK
www.dk.com

NOTE TO PARENTS AND TEACHERS

The **Eyewitness Guides** series encourages children to observe and question the world around them. It will help families to answer their questions about why and how things work – from daily occurrences in the home to the mysteries of space. By regularly "looking things up" in these books, parents can promote reading for information every day.

At school, these books are a valuable resource. Teachers will find them especially useful for topic work in many subjects and can use the experiments and demonstrations in the books as inspiration for classroom activities and projects. **Eyewitness Guides** titles are also ideal reference books, providing a wealth of information about all areas of science within the curriculum.

Project Editor Sharon Lucas
Designer Heather McCarry
DTP Manager Joanna Figg-Latham
Production Eunice Paterson
Managing Editor Josephine Buchanan
Senior Art Editor Neville Graham
Special Photography Dave King
Editorial Consultant Alan Morton,
Science Museum, London
Special Consultant Jack Challoner

This Eyewitness ®/™ Science book
first published in Great Britain in 1992 by
Dorling Kindersley Limited, 9 Henrietta Street,
London WC2E 8PS

2 4 6 8 10 9 7 5 3

Copyright © 1992 Dorling Kindersley Limited, London

A CIP catalogue record for this book is available
from the British Library.

ISBN 0 7513 6135 6

Reproduced by Colourscan, Singapore
Printed in China by
Toppan Printing Co., (Shenzhen) Ltd

Thermometer of
Lyons (18th century)

Bunsen
burner
(1872)

Ancient
Egyptian
mirror

Items from a box of flame test
equipment (19th century)

Contents

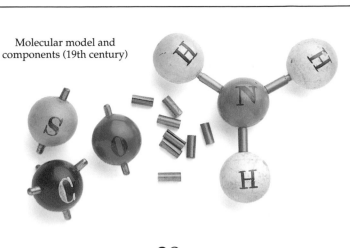

Molecular model and
components (19th century)

What is matter?

Everything found everywhere in the Universe – from the farthest star to the smallest speck of dust – is made of matter in an incredible variety of forms. About 200 years ago heat was regarded by many scientists as being a special sort of matter. But now it is known that heat is simply the motion of tiny particles of matter (pp. 38-39). Sound, too, is a certain type of movement of matter. Forms of energy such as radiation (for example, light, radio waves, and X-rays) are generally regarded as not being matter, though they are very closely linked to it. All the different kinds of matter have one thing in common – mass. This is the amount of material in any object, and shows itself as resistance to being moved. A truck, for example, has more mass and is much harder to move than a toy car. Every piece of matter in the Universe attracts every other piece of matter. The amount of matter is important – a large piece attracts other matter more strongly than a small piece.

A CONTAINED UNIVERSE
This terrarium is a microcosm of the living world. It contains the three states of matter – solids (pp. 12-13), liquids (pp. 18-19), and gases (pp. 20-21), as well as interesting substances found in the world of matter.

THE LIVING WORLD
All living matter (pp. 42-43) can organize itself into intricate forms and behave in complicated ways. It was once thought that matter in living things was controlled by a "vital principle", a sort of ghostly force. But now scientists think that living and non-living matter obey the same laws.

*Plants stretch upwards
to reach the light*

MIXING AND SEPARATING MATTER
Gravel, sand, and water can be made into a mixture (pp. 26-27), and can easily be separated afterwards. Each of these materials is made of other substances that are more strongly combined, and very hard to separate. Water, for example, is a combination of the gases hydrogen and oxygen. Such a close combination is called a chemical compound (pp. 26-27).

*A mixture of gravel,
sand, and water*

METALLIC MATTER
Metals (pp. 16-17) are found in rocks, called ores. Pure metals are rare, and usually they have to be separated from their ores. Once separated, they are often combined with other materials to form alloys – mixtures of metals and other substances.

*Lead is a metal which looks
solid, but flows extremely
slowly over decades*

SOLUTIONS AND COLLOIDS

Substances can often dissolve in a liquid or solid. They form solutions – they are mixed very thoroughly with the liquid or solid, breaking up into groups of a few atoms, or even into single atoms (these are the smallest normally existing particles of matter, pp. 34-35). A colloid (pp. 24-25) consists of larger particles of matter that are suspended in a solid, liquid, or gas.

Glass is transparent matter

THE WORLD OF GASES

When the particles of a substance become separated from each other, the substance becomes a gas. It has no shape of its own, but expands to fill any space available. Air (largely a mixture of nitrogen and oxygen) was the first gas to be recognized. It was many centuries before scientists realized that there are other gases as well as air. This was because gases tend to look similar – they are mostly colourless and transparent.

Condensation is caused by molecules of water vapour cooling and turning into a liquid

LIQUID MATTER

Liquids, like gases, consist of matter that can flow, but unlike gases, settle at the bottom of any container. Nearly all substances are liquids at certain temperatures. The most important liquid for living creatures is water. Most of the human body is made up of water. It forms the bulk of human blood, which transports dissolved foodstuffs and waste products around the body.

This butterfly is made of some of the millions of varieties of living matter on Earth

Water contains dissolved oxygen and carbon dioxide gases from the air

SOLID SHAPES

The metal and glass (pp. 24-25) of this terrarium could not act as a container for plant and animal specimens if they did not keep a constant shape. Matter that keeps a definite shape is called solid. However, most solids will lose their shape if they are heated sufficiently, turning into a liquid or a gas.

Solids such as rock keep a definite shape

Ideas of the Greeks

Ancient Greek philosophers vigorously debated the nature of matter, and concluded that behind its apparent complexity the world was really very simple. Thales (about 600 BC) suggested that all matter was made of water. Empedocles (5th century BC) believed that all matter consisted of four basic substances, or elements – earth, water, air, and fire – mixed in various proportions. In the next century Aristotle added a fifth element from the heavens – the ether. Leucippus (5th century BC) had another theory that there was just one kind of matter. He thought that if matter was repeatedly cut up, the end result would be an uncuttable piece of matter. His follower Democritus (about 400 BC) called these indivisible pieces of matter "atoms" (pp. 34-35), meaning "uncuttable". But Aristotle, who did not believe in atoms, was the most influential philosopher for the next 2,000 years, and his ideas about elements prevailed.

THE FOUR ELEMENTS IN A LOG
Empedocles' idea of the four elements was linked to certain properties. Earth was dry and cold, water was wet and cold, fire was hot and dry, and air was hot and wet. In the burning log below, all four elements can be seen. Empedocles thought that when one substance changes into another – such as when a burning log gives off smoke, emits sap, and produces ash – the elements that make up the log are separating or recombining under the influence of two forces. These forces were love (the combining force), and hate or discord (the separating force).

Empedocles

Design on the coin has been smoothed away

WEARING SMOOTH
Ancient philosophers thought that when objects such as coins and statues wore smooth with the passing of time they were losing tiny, invisible particles of matter.

MODEL OF WATER
Plato (4th century BC) thought that water was made up of icosahedrons, a solid shape with 20 triangular faces.

LIQUID FROM A LOG
Empedocles believed that all liquids, even thick ones like the sap that oozes from a burning log, were mainly water. His theory also said that small amounts of other elements would always be mixed with the main element.

Sap is made from the element water

Ash and cinders are mainly the element earth

ELEMENTAL ATOMS
Democritus developed the theory of atoms and combined it with the theory of elements. Like Plato, he thought there were only four shapes of atom, one for each element. He argued against the religious beliefs of his day, claiming that atoms moved randomly, and that there were no gods controlling the Universe.

MODEL OF EARTH
Atoms of earth were thought by Plato to be cubes, which can stack tightly together to give strength and solidity.

ASHES TO ASHES
The theory of the elements suggested that ash and cinders were mainly made of the element earth, with a little of the element fire. At the end of the burning process, there was not enough fire left for more ash to be produced, but some fire remained for a while in the form of heat. The Greeks thought that the elements earth and water had a natural tendency to fall.

THE FIVE ELEMENTS
The human figure in this engraving is standing on two globes, representing earth and water, and is holding air and fire in his hands. The sun, moon, and stars are made of the ether, the fifth element.

MODEL OF AIR
Plato's model of an air atom was an octahedron, a solid figure with eight faces.

Smoke is mostly air, with some earth in the form of soot mixed in

NO SMOKE WITHOUT FIRE
When a piece of matter is burned, the element air inside was thought to be released in the form of smoke. The Greeks thought that air, like fire, had a natural tendency to rise.

MODEL OF FIRE
According to Plato, the atom of fire was a solid shape with four sides called a tetrahedron.

PENETRATING FLAMES
The element fire could be seen most clearly in flames and sparks, but the Greeks thought that some fire was present in everything. Plato's model of the fire atom is sharp and pointed. This is because heat seemed to be able to penetrate virtually every piece of matter.

Flames and sparks are the element fire

Investigating matter

IDEAS ABOUT MATTER AND HOW IT BEHAVED changed little for hundreds of years. But in Europe during the 16th and 17th centuries "natural philosophers" looked again at the ancient theories about matter. They tested them, together with newer ideas about how matter behaved, by experiments and investigations, and used the newly invented microscope and telescope to look closely at matter. Measurements became more precise. News of discoveries was spread by the printing press. The scientific revolution had begun.

SANDS OF TIME
The sandglass was a simple device, which allowed scientists to work out how fast objects fell, or how long it took for chemicals to react. More accurate measurements of time were not possible until after the first pendulum clock was made in 1657.

Flow of sand is regulated by the narrow glass channel

LABOURING IN THE LAB
This 17th-century laboratory illustrates just some of the processes used by "natural philosophers" to find out about matter.

Eyepiece

HEATING AND COOLING MATTER
Very early experiments showed the effects of heating or cooling of matter. Philo of Byzantium constructed his lead thermoscope (Greek for "observing heat") about 250 BC. When the globe on the right is warmed, the air inside expands and pushes its way up the tube, which is immersed in the water on the left. If the heat is strong enough, bubbles of air escape. When the globe is cooled, the air contracts and water is drawn back up the tube.

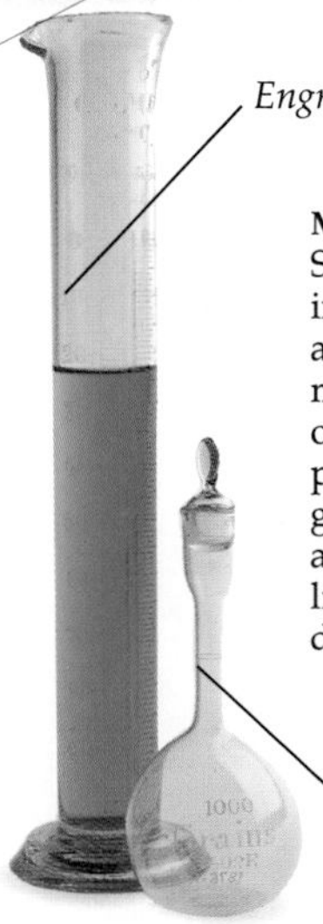

Reconstruction of Philo's thermoscope

Tiny bubbles of air

Lead globe is full of air

Glass globe contains water

Engraved scale

MARKINGS FOR MEASURING
Scientific investigations often involved measuring the exact amount of a liquid. This tall measuring cylinder has a scale of accurate markings for this purpose, while the specific gravity bottle has just one very accurate marking. When the liquid inside is weighed, its density can be calculated.

Single accurate marking

Object to be viewed is placed on the glass

Small lens focuses light

Tilting mirror

SMALL WONDER
Microscopes began to open up the world of the very small from the mid 1500s onwards. In the mid 1600s Anton van Leeuwenhoek found that a single drop of pond water could contain 8 million "animalcules" – tiny but intricately constructed creatures and plants. The more elaborate microscope shown here was made by Edmund Culpeper of London, in about 1728. It used a tilting mirror, visible at the bottom, to reflect light on to a specimen mounted above it on glass.

Enlarged view of a deathwatch beetle

DEGREES OF ACCURACY

As scientists looked closer at matter, they needed ever more accurate ways of measuring what they saw. This thermometer, a device for measuring changes in temperature, was made in Florence, Italy, in the 18th century. The bulb at the bottom contained alcohol, which expanded when it became warmer and moved along the coiled tube. The tube is marked with dots at equal intervals.

Dots are regularly spaced

Bulb of alcohol

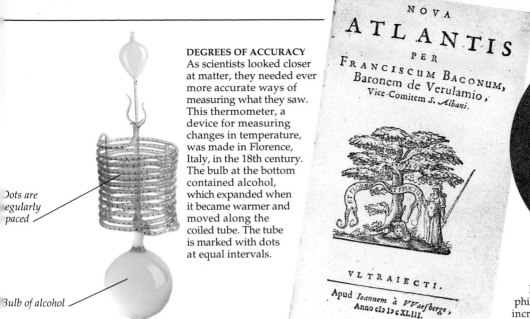

SCIENTIFIC IMPROVEMENTS

Francis Bacon (1561-1626), the English philosopher, hoped that the new science could increase human well-being. *The New Atlantis* was his 1626 account of an imaginary society, or "utopia", where the government organized teams of scientists to conduct research and use the results to improve industry.

Ivory arm

Brass ring

Pointer indicates equilibrium

Cord fulcrums

Alchemy

Before the scientific revolution of the 17th century, the closest approach to a systematic study of matter was alchemy. This was enthusiastically studied in Egypt, China, and India at least as early as the 2nd century BC, and from the Middle East it eventually reached Europe. Alchemists learned much from the practical skills of dyers and metalworkers, and borrowed various ideas from astrologers. They tried, without success, to change "base" metals, such as lead, into precious ones, such as silver or gold. This series of operations was described as "killing" the metal and then "reviving" it. Alchemists also attempted to make the elixir of life, a potion that would give them the secret of everlasting life.

Spout-like alembic sits on the cucurbit

"Fool's gold" – a compound made up of iron and sulphur

QUEST FOR GOLD

Alchemists used all kinds of scientific instruments and chemical processes in their mystical quest for gold. The laboratory above was imagined by a 19th-century painter.

Alembic and cucurbit

PURE MATTER

Cucurbits and alembics were used by many alchemists to purify liquids. As the cucurbit was heated, vapour from the liquid inside rose to the top, then cooled and condensed. The pure liquid dripped from the alembic and was then collected.

Compound counterbalance

HANGING IN THE BALANCE

Scales are one of the most basic of measuring instruments. The weight at the right of these Chinese scales is moved along the longer arm until it balances the object in the pan. This particular method is quick, convenient, and fairly accurate. It was not until the 17th century that chemists realized that accurately weighing the substances involved in a chemical reaction is crucial to understanding what is going on.

Greek lead weight

Solid matter

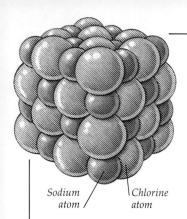

Sodium atom / **Chlorine atom**

HELD IN PLACE
As in most solids, the atoms (pp. 34-35) in this model of salt hold each other firmly in place, and form a regular pattern.

Ever since people began to observe the world carefully, they have classified matter into three main states – solids, liquids (pp. 18-19), and gases (pp. 20-21). A piece of solid matter has a fairly definite shape, unlike a liquid or a gas. Changing the shape of a solid always requires a certain amount of force, which can be either large or small. Squeezing or stretching a solid can change its volume (the amount of space it takes up) but generally not by very much. When they are heated, most solids will turn to liquid, then to gas as they reach higher temperatures. However, some solids such as limestone (pp. 36-37) decompose when they are heated. Crystals (pp. 14-15) and metals (pp. 16-17) are two of the most important kinds of solid.

Screw

Gimbal ring keeps the compass level even when the ship is rolling

SOLID STRENGTH
Brass, which is a combination of metals, or an alloy (pp. 16-17), is made of copper and zinc. It is used for the compass's gimbal ring, mounting ring, and pivot. Brass is strong, so the gimbal ring's mountings will not quickly wear out. Like many metals, brass is not magnetic, and will not interfere with the working of the compass needle.

Mounting ring fixes the compass in its case and holds it under the glass

SOLID PROPERTIES
Like most artificial objects, this 19th-century mariner's compass contains several kinds of solid. The compass has been taken apart on these two pages to reveal four solids – metal, card, wood, and glass. All the solids in the compass have been chosen for their special and varied properties.

Mariner's compass in its protective wooden box

Magnet

Pins are attracted to the ends of the magnet

MAKING MAGNETS
The ancient Chinese are thought to be the first magnet makers. They discovered that iron can be magnetized by making it red-hot and then letting it cool while aligned in a north-south direction.

Hole which rests on pointed pivot

FINDING THE WAY
Beneath the compass card is a magnet, made either of iron or of a rock called lodestone. Magnets attract or repel each other, and also respond to the Earth's magnetic poles. They tend to swing into a north-south line if they are free to move. The compass showed the mariner the angle between the direction of the ship and the north-south direction of the magnet.

SIZE AND STRENGTH
Galileo Galilei (1564-1642) studied the strength of materials and showed that there is a limit to land animals' size. If the largest dinosaur doubled in size, its bones would become larger and stronger. However, the increase in the dinosaur's weight would be even greater, making its bones snap.

AT FULL STRETCH
Many solids are elastic – after being stretched or squeezed, they return to their original shape. A rubber band, for example, can be stretched to more than double its length and then return to its original length. But if a material is distorted too much its shape may be permanently altered.

Pointed pivot supports the compass

Brass knobs sit in holes in the gimbal ring

Glass flows over hundreds of years

SEEING THROUGH THINGS
The front of the compass needs to be transparent and strong. It is made of glass – halfway between a solid and a liquid (pp. 24-25). Glass may seem rigid, but over hundreds of years it gradually flows and becomes distorted. Most solids block light completely, but the clearest kinds of glass absorb little of the light passing through them.

QUALITIES OF WOOD
The protective container of the compass needs to be strong, and to be rigid (keep its shape). Wood has many different qualities – the wood used for this container is fairly hard and long-lasting. Yet it is also soft and light enough to be easily worked with metal tools, and can be carved to form a smooth bowl shape.

DIFFERENT VIEWS
Transparent (see-through) materials can give a clear and undistorted view, as in the glass front of a watch. Or they can be deliberately shaped to help give an even clearer view, as in eyeglasses.

Diamond
10

Considerable difference in hardness between diamond and the other minerals on the scale

FROM SOFT TO HARD
Scientists classify solid materials by their hardness, on a scale from one to ten named after Friedrich Mohs (1773-1839). These solids are all minerals (so-called because they are usually mined). Talc is the softest at one, and diamond the hardest at ten. Any solid on the scale will scratch a softer one, and will itself be scratched by a harder one.

Compass direction points

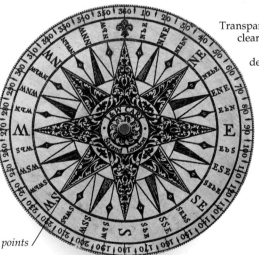

Corundum
Topaz
Quartz
Feldspar
Apatite
Fluorite
Calcite
Gypsum
Talc

1 2 3 4 5 6 7 8 9

PAPER POINTERS
The compass points are printed on paper or card. Paper is made by pulping up wood and treating it to make it soft and flexible. It consists of countless fibres, and absorbs ink well because the ink lies in the spaces between the fibres.

The world of crystals

CRYSTALS HAVE BEEN VIEWED with awe since ancient times. They are often very beautiful, and their shapes can differ widely, yet all crystal forms are of just six basic kinds. The orderly shape of each crystal is created by the arrangement of the atoms (pp. 34-35) inside. With the help of powerful microscopes, many objects and materials that seem irregularly shaped to the naked eye, such as stalactites and most metals, can be seen to be masses of tiny uniform crystals. Many crystals are valuable in industry, and some, such as quartz (used in watches) and silicon (used in computers), can be made in the laboratory.

SIX SHAPES
Abbé René Haüy (1743-1822) was one of the first to show that crystal shapes fall into six geometrical groups. He suggested how they could be built up by stacking identical units in regular patterns.

SCULPTED WATER
Stalactites are mainly limestone, created by centuries of dripping water. The atoms in the limestone have arranged themselves in regular crystalline patterns.

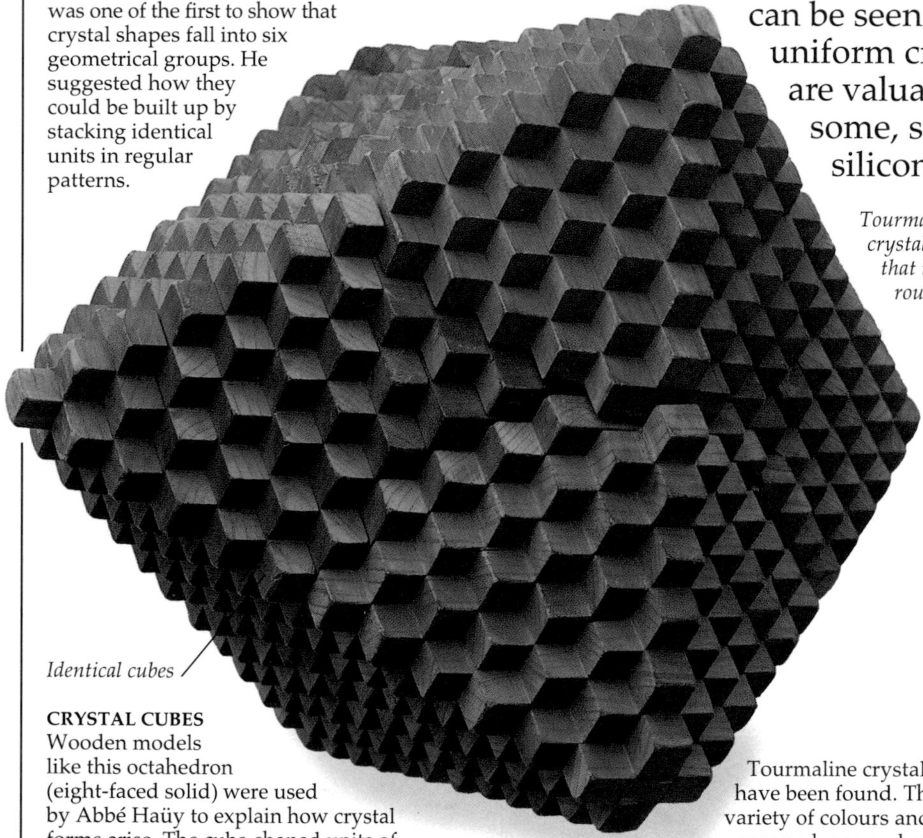

Tourmaline forms fine, long crystals with a cross-section that is triangular with rounded corners

Identical cubes

CRYSTAL CUBES
Wooden models like this octahedron (eight-faced solid) were used by Abbé Haüy to explain how crystal forms arise. The cube-shaped units of this crystal model are arranged in square layers, each larger than the previous one by an extra "border" of cubes.

HIGH-RISE CRYSTALS
Tourmaline crystals up to 3 m (10 ft) long have been found. They can occur in a wide variety of colours and are prized as gems. If warmed, one end of a tourmaline crystal becomes positively charged, and the other end becomes negatively charged.

THE EMERALD CITY
Crystals are often used as symbols of perfection and power. The magical Emerald City appears in the 1939 film *The Wizard of Oz*.

Outer part of the bismuth cooled fast and formed only microscopic crystals

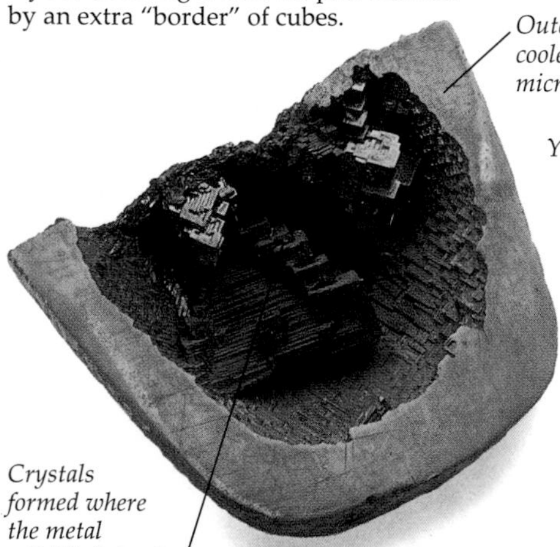

Yellow sulphur crystals

Needle-shaped crystals formed where alloy solidified slowly

Crystals formed where the metal solidified slowly

BOX-LIKE BISMUTH
Inside this piece of bismuth are intricate "nests" of crystal boxes, formed as the metal slowly solidified.

MELLOW YELLOW
At low temperatures, fairly flat sulphur crystals form. At high temperatures they are needle-shaped.

METAL MIXTURE
The thin, pointed crystals shown here are alloys (pp. 16-17) of copper and aluminium.

Outer part of alloy cooled fast and formed few crystals

AMAZING ARAGONITE

Crystals of aragonite can be found in limestone caves and hot springs. They take many forms, such as fibres, columns, or needles. Their colour is usually white, yellow, green, or blue.

Aragonite often forms twin crystals

EGG-SHAPED ATOMS

In this model made by Wollaston, the atoms in the crystal are imagined to be egg-shaped. Each has six neighbours at the sides, forming a strong horizontal layer.

MAKING CRYSTALS CLEAR

William Hyde Wollaston (1766-1828) made important contributions to crystallography (the scientific study of crystals). He recognized that a cubic crystal, for example, did not have to be built from cubes. Instead, it could be assembled from atoms of other shapes, as in his wooden models shown on this page. Scientists now know that atoms can take very complicated shapes when they join together.

LOOSE LAYERS

If atoms in crystals are spheres, Wollaston realized that they would have neighbours on all sides. They would not form such strong layers as the atoms of the other shapes shown here.

FLAT ATOMS

Wollaston thought that if crystal atoms are flat, they would link most strongly where their flat faces were in contact. They might form columns or fibres.

BLUE IS THE COLOUR

The mineral azurite is blue, as its name implies. In the past azurite was crushed and used as a pigment. It contains copper and is found with deposits of copper ore. When azurite is made into a gem, it can be faceted – cut to display polished flat faces.

Azurite crystals are "knobbly"

LINING UP LIQUID CRYSTALS

Some crystals are liquid. The particles in a liquid can be temporarily lined up in regular arrays when an electric or magnetic field is applied. The liquid crystals shown here were revealed under an electron microscope. The liquid affects light differently when the crystals form, and can change from being transparent to being opaque, or coloured. In digital clocks and watches, calculators, or laptop computers, electricity is used to alter segments of the display from clear to dark, to generate the changing numbers or letters.

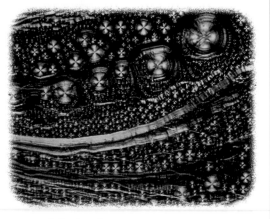

Metals and alloys

THE THREE METALS most widely used are iron, steel, and aluminium. Iron and aluminium are both metallic elements (pp. 32-33), but steel is a mixture of iron and carbon. Such a combination, either of metals, or of metals and non-metals, is called an alloy. By combining a metal with other substances (either metallic or non-metallic) it can often be made stronger. Most metals are found in ore (rock), combined with other elements such as oxygen and sulphur. Heating the ore separates and purifies the metal. When metals are pure they are shiny, can be beaten into shape, and drawn out into wires. They are not brittle, but are often rather soft. Metals are good conductors (carriers) of electric current and heat.

Surface pitted by heating during fall

Rust formed by iron combining with oxygen from the air after the fall

GOLDEN CHARACTER
Gold is a precious metal. It is rare and does not tarnish. It can be beaten into sheets of gold leaf, which are used to decorate letters in illuminated manuscripts like the one above.

HEAVENLY METAL
A very pure iron comes from certain meteorites. They are bodies that have fallen to Earth from outer space and have been partly burned away by the friction of entering the Earth's atmosphere.

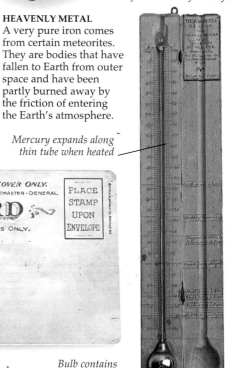

Mercury expands along thin tube when heated

CUTTING EDGE
The use of bronze dates from about 5000 BC in the Middle East, and 2000 BC in Europe. Bronze is an alloy of copper and tin. It is very hard, and was used for the blades of axes, daggers, swords, and razors.

Blade of hammered bronze

Ancient Egyptian razor

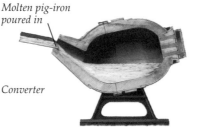

Molten pig-iron poured in

Bronze handle

HARD WORDS
Aluminium makes up one-twelfth of the rock near the Earth's surface. It was discovered in 1809, but only came into widespread use after 1886. It is a very light metal, and was first used in jewellery and novelties, like this picture postcard. Aircraft parts are often made of aluminium alloys.

Bulb contains mercury

LIQUID METAL
The metal mercury, sometimes known as quicksilver, is a liquid at normal temperatures. It expands by a relatively large amount when warmed, and has long been used in instruments to measure temperature, such as this 18th-century thermometer.

Henry Bessemer

Converter

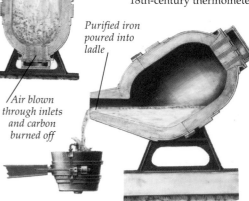

Purified iron poured into ladle

Air blown through inlets and carbon burned off

MAN OF STEEL
Henry Bessemer (1813-1898) greatly speeded up the steel-making process in the mid-19th century with his famous converter. Air was blown through molten pig-iron (iron ore that had been heated in a furnace with coal or wood). This burned off the carbon (pp. 40-41) from the coal or wood in a fountain of sparks. The purified iron, still molten, was tipped out of the converter, and measured amounts of carbon and metals, such as nickel, manganese, or chromium, were added. These other substances turned the molten iron into steel, an alloy that is renowned for its strength.

BELL OF BRONZE

Metals such as bronze are ideal for bells because they vibrate for a long time after being struck. From about 1000 BC bronze has been cast (poured in a molten state into a mould). Once cast, large bells must cool very slowly to prevent cracking. The Liberty Bell, which hangs in Philadelphia, Pennsylvania, USA, weighs about 943 kg (2,079 lb), and is about 1 metre (3 ft) high. It was made in London and delivered in 1752, but it cracked and had to be recast twice before it was hung. It cracked again in 1835 and in 1846. Since then it has never been rung.

PURE QUALITY

Goldsmiths formerly judged gold's purity by scraping it on a type of dark rock called touchstone. Its streak was then compared with the streaks made by the gold samples on two "stars". The best match came from gold of the same purity.

Sample has 875 parts of gold per thousand

Sample has 125 parts of gold per thousand

Streaks left by scraping samples on touchstone

Steel spring

Steel wheels

MULTIPURPOSE METALS

Various metals, each with its own particular job to do, are used in this old clock. The springs, chain, and cogwheels inside, which receive the most wear, are made of steel. The case is made of brass, an alloy of copper and zinc that is not so strong as steel. To make it look attractive, the brass has been gilded (coated with gold).

Cross-section of transatlantic cable

Rubber-like gutta-percha prevents electricity from leaking

Twisted steel cables

DEEP COMMUNICATION

A telegraph cable 3,740 km (2,325 miles) long was first laid across the sea-bed of the Atlantic Ocean in 1850, linking Britain with the US. The outer part of a typical cable was a strong sheath of twisted steel cables. It would resist rusting even in seawater (a solution known to rust metals quickly).

Twisted steel cables

Copper strands

OCEAN CURRENT

A seven-stranded copper wire lies at the heart of the undersea cable. Copper was chosen for the wire because of its very useful properties. It is an excellent carrier of electric current and is easily drawn out into wires.

Properties of liquids

ACCORDING TO THE GREEKS who believed in the four elements, all liquids contain a large proportion of water (pp. 8-9). However, the Greeks who believed in atoms (pp. 34-35) thought that the atoms in a liquid could slide around each other, making the liquid flow to take the shape of its container. This is also the modern view. Liquid particles attract each other and keep close together, so they cannot be easily squeezed into a smaller volume or stretched out into a larger one. When a liquid is heated, however, the spacing between the particles generally increases in size, so the liquid expands. When a liquid is cooled, the reverse effect occurs, and the liquid contracts. It is possible for liquids to dissolve some solid substances. For example, salt placed in water seems to disappear very slowly. In fact the salt breaks up into individual atoms of sodium and chlorine (pp. 50-51). The ions spread out through the water, forming a mixture (pp. 26-27) called a solution of salt in water. Liquids can also dissolve gases and other liquids.

Hydrogen atom

Oxygen atom

SLIPPING AND SLIDING
The smallest unit of water consists of one atom (pp. 34-35) of oxygen joined to two hydrogen atoms. These clusters of three atoms slide around each other in liquid water.

PRESSING MATTER
Most liquids, particularly water and oil, act as good transmitters of pressure. In 1795 Joseph Bramah (1749-1814) patented his hydraulic press, in which compressed fluid multiplied the force that could be exerted by a human operator.

SLOW FLOW
Some liquids flow easily, but honey flows very slowly and is described as being "viscous". Liquids such as tar and pitch (a substance used to seal roofs) are even more viscous.

Slow-moving liquid

Meniscus

Level surface

Droplets are forced into shape by surface tension

Gas in a liquid is pressed into spherical or almost spherical bubble shapes by the surrounding liquid

Liquid spreads out in a thin film

CURVED EDGE
The surface of an undisturbed liquid is horizontal, except at the very edge, where it forms a curve called a meniscus. The meniscus can be curved up as it is here, or curved down.

UPLIFTING MATERIAL
Because a liquid can flow, an object can be pushed down into it, forcing some of the liquid aside. But the displaced liquid tries to flow back, pushing the object upward. The object then seems lighter than the liquid, and can even float, like this boat.

WORN DOWN BY WATER
Given sufficient time, flowing liquids wear away solid surfaces, even rocks. The abrasive effect is increased when the liquid carries solid particles of rock and mud. Some rocks, such as clays and sandstones, have a low resistance to erosion. This canyon in the Arizona desert has been worn away by 10,000 years of flash floods.

Narrow neck of vessel causes liquid to speed up as it flows through

Fast-moving liquid

Liquid takes the shape of its container

THE STRENGTH OF MOVING LIQUIDS
A stream of liquid can deliver a powerful force – a tsunami, or "tidal wave", can sweep away towns. Slower-moving liquid has time to break up and flow round obstacles, and does them less damage. Where a liquid has escaped from a container, surface tension (the inward pull at the surface) tries to pull it into shape. But because it is a relatively weak force, surface tension can pull only small amounts into drops – larger quantities of liquid are chaotic and formless.

WATER POWER
Streams and rivers have been used to turn waterwheels since ancient times. In 18th-century Britain water-powered looms featured heavily in the Industrial Revolution. Today, water from lakes, reservoirs, or the sea is used to turn electricity-generating turbines worldwide.

Level surface

BRIMMING OVER
The tiny particles that make up a liquid are held together by their attraction to each other. Surface tension makes the surface of a liquid behave like the tight elastic skin of a balloon. The wine in this glass is above the brim, but surface tension stops it from overflowing.

WALKING ON WATER
Surface tension lets this insect's feet make dents in the water, but they don't go through the "skin" on the surface.

Gases and their properties

Aɴᴄɪᴇɴᴛ ᴘʜɪʟᴏsᴏᴘʜᴇʀs ᴡᴇʀᴇ ᴘᴜᴢᴢʟᴇᴅ by the exact nature of gases. They realized air was not empty space. Some guessed the smell of a perfume was due to the spreading of tiny particles, and that frost was formed by the condensation of invisible water vapour. Many observed that the wind bent trees and vigorous bubbling made water froth. These early philosophers believed there was a single element of air (pp. 8-9), which had "levity", a tendency to rise. In the 17th century Evangelista Torricelli (1608-1647) showed that air, like solids and liquids, can be weighed. In the next century, chemists showed that air is a mixture of gases, and identified the gases given off in chemical reactions. These newly discovered gases were soon put to use; for example, gas obtained from coal produced light and heat.

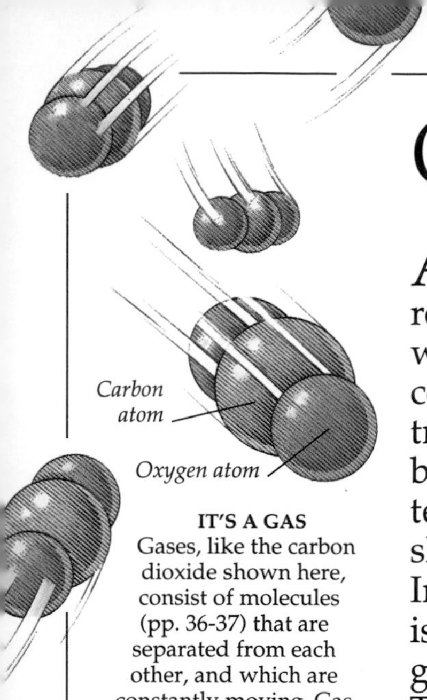

Carbon atom

Oxygen atom

IT'S A GAS
Gases, like the carbon dioxide shown here, consist of molecules (pp. 36-37) that are separated from each other, and which are constantly moving. Gas molecules are usually complex, made up of atoms (pp. 34-35) that are closely bound together.

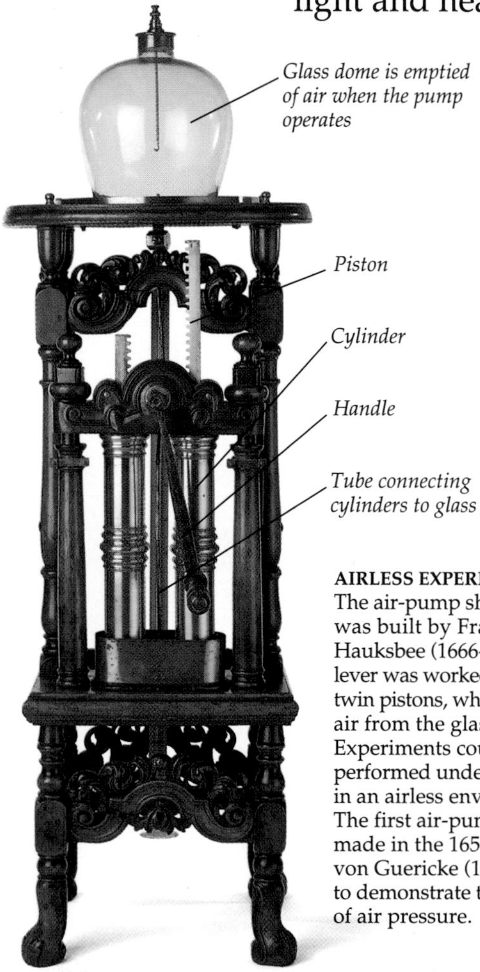

Glass dome is emptied of air when the pump operates

Bung

Oxygen travels down the tube

Heating the potassium permanganate crystals gives off oxygen

Piston

Cylinder

Handle

Tube connecting cylinders to glass dome

Test tube stand

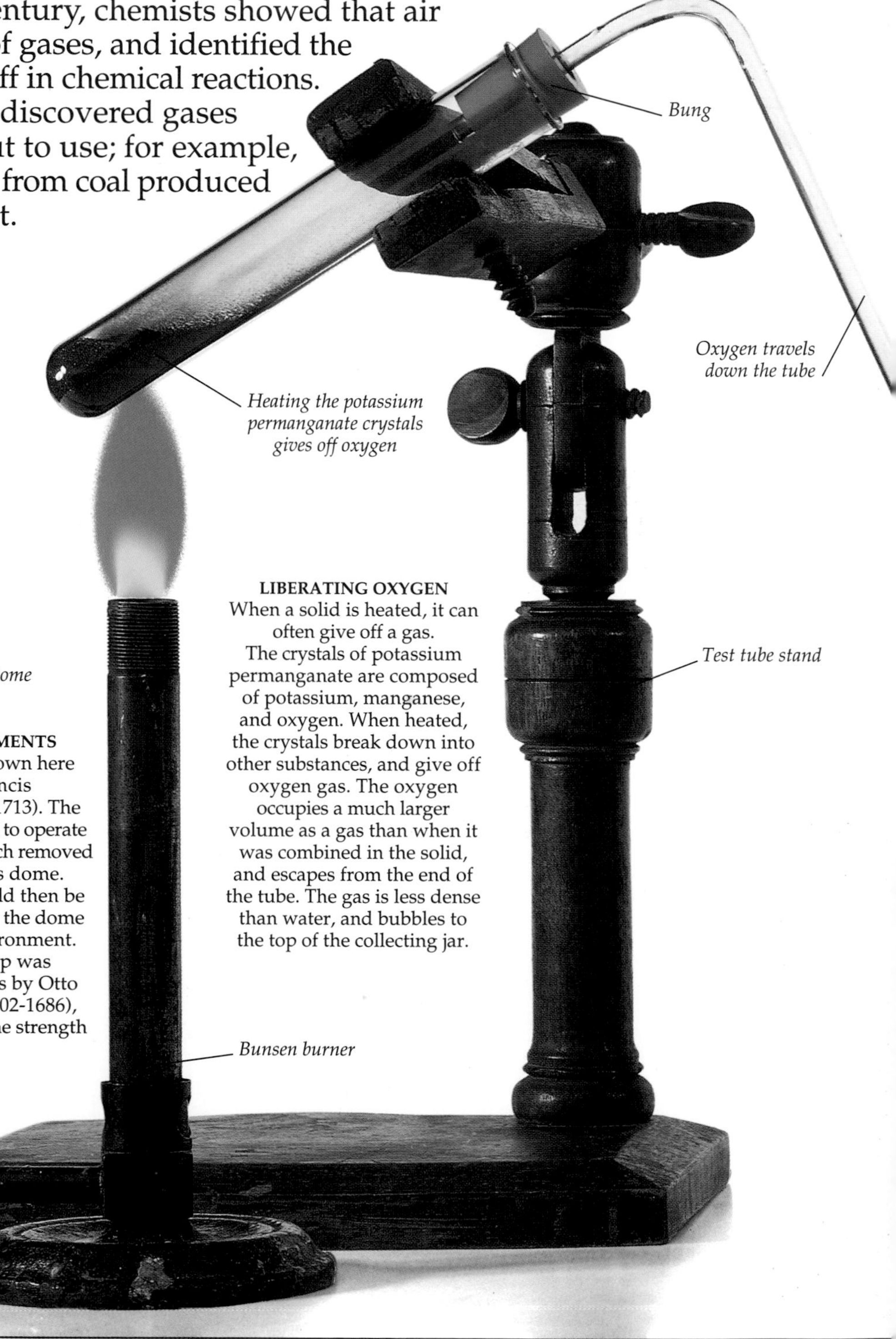

LIBERATING OXYGEN
When a solid is heated, it can often give off a gas. The crystals of potassium permanganate are composed of potassium, manganese, and oxygen. When heated, the crystals break down into other substances, and give off oxygen gas. The oxygen occupies a much larger volume as a gas than when it was combined in the solid, and escapes from the end of the tube. The gas is less dense than water, and bubbles to the top of the collecting jar.

AIRLESS EXPERIMENTS
The air-pump shown here was built by Francis Hauksbee (1666-1713). The lever was worked to operate twin pistons, which removed air from the glass dome. Experiments could then be performed under the dome in an airless environment. The first air-pump was made in the 1650s by Otto von Guericke (1602-1686), to demonstrate the strength of air pressure.

Bunsen burner

Gas supply

PRESSING FORCE
A barometer measures changes in atmospheric pressure. The early barometer made by Evangelista Torricelli had a mercury-filled upright glass tube. The tube's open end dipped in a bowl of mercury. Atmospheric pressure forced the mercury down in the bowl, and balanced the weight of the mercury in the tube.

A VALUABLE COLLECTION
This apparatus was used by Joseph Priestley (1733-1804) to collect gases. The gases would bubble up through the water and be collected for experimentation in the glass jars.

HIGH ACHIEVER
Jacques Charles (1746-1823) discovered an important law about the expansion of gases when heated (p. 39). In 1783 he took part in the first flight in a hydrogen balloon. Hydrogen is very light and also highly flammable, but was still being used in airships in the 1930s.

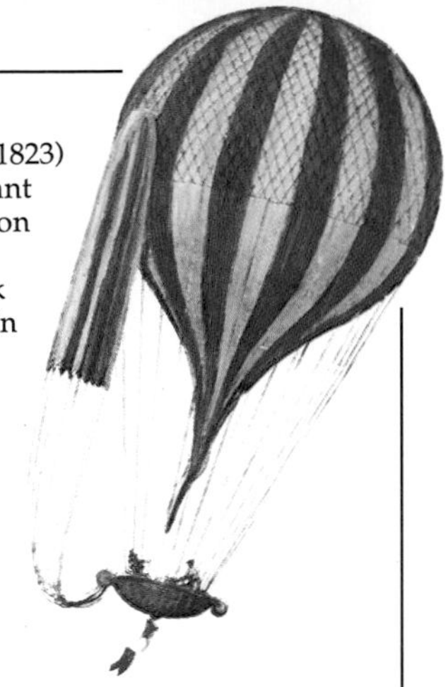

GAS WORKER
In 1775 Joseph Priestley discovered oxygen during his work with mercuric oxide. He found that oxygen supported breathing and burning, but did not recognize its exact nature. Using carbon dioxide from a local brewer to make water fizzy, he invented soda-water.

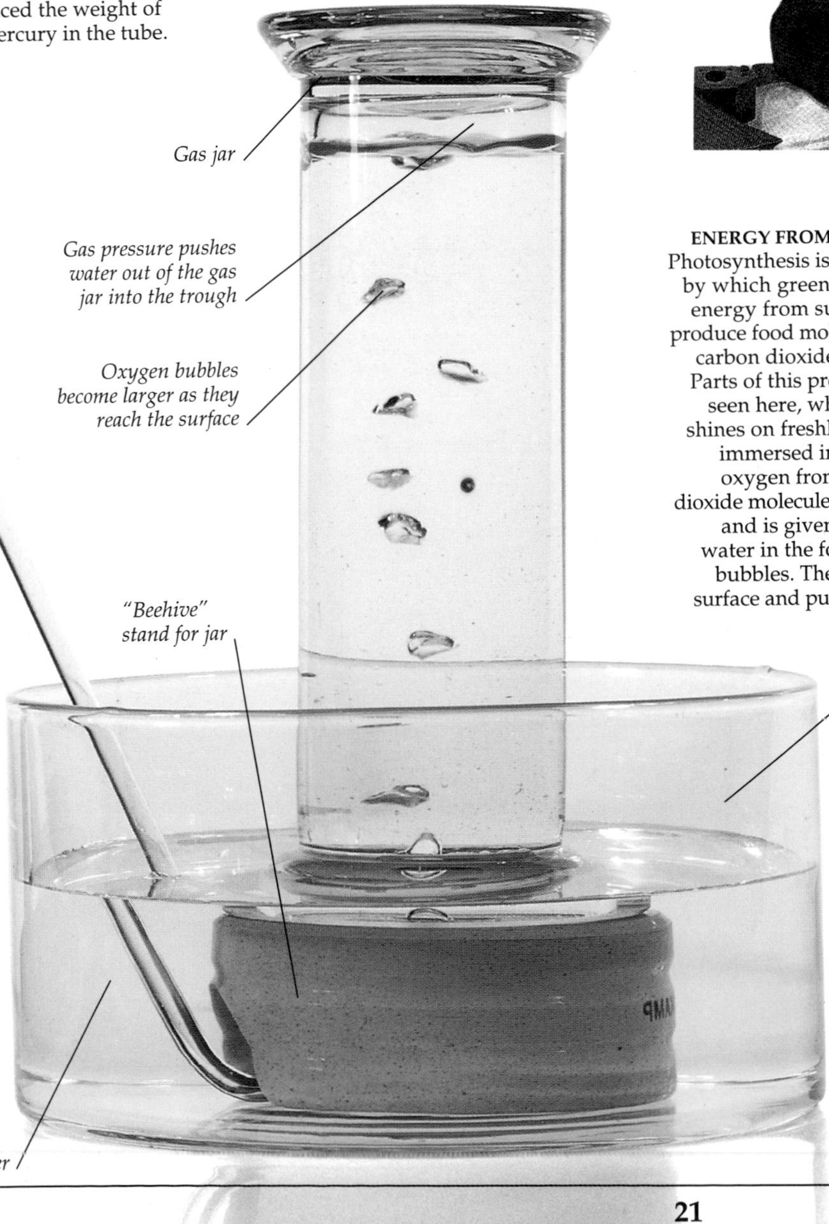

Gas jar

Gas pressure pushes water out of the gas jar into the trough

Oxygen bubbles become larger as they reach the surface

"Beehive" stand for jar

Water

ENERGY FROM SUNLIGHT
Photosynthesis is the process by which green plants take energy from sunlight, and produce food molecules from carbon dioxide and water. Parts of this process can be seen here, when sunlight shines on freshly cut leaves immersed in water. The oxygen from the carbon dioxide molecules is released, and is given off into the water in the form of small bubbles. They rise to the surface and push water out from the jar.

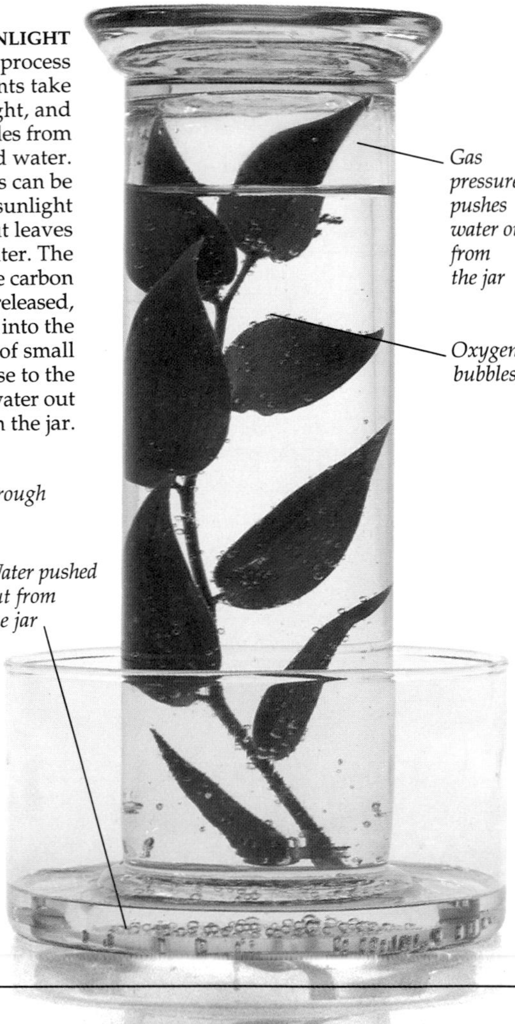

Trough

Gas pressure pushes water out from the jar

Oxygen bubbles

Water pushed out from the jar

Changes of state

MATTER CAN BE ALTERED in various ways. Heating a solid to a temperature called its melting point will make it change state – it will turn into a liquid. Heating a liquid to a temperature called its boiling point has a similar effect – the liquid will change state and turn into a gas. It is possible to affect both the melting and the boiling point of matter. For example, if an impurity such as salt is added to ice, the melting point of the ice is lowered. The mixture (pp. 26-27) of salt and ice will melt, while pure ice at the same temperature will remain frozen. If salt is added to water, it raises the boiling point, and the water boils at a higher temperature. Pressure can also affect the state of matter. Where air pressure is low, the boiling point of water is lowered. An increase in pressure will lower the melting point of a solid.

YIELDING TO PRESSURE
Where a wire presses on ice, the melting point is lowered, and the ice melts. The wire cuts through the ice, which freezes again as the wire passes.

FROM SOLID TO GAS
In the following sequence, heating a solid – ice – to its melting point makes it change state to a liquid. Heating the liquid – water – to its boiling point makes it change state to a gas.

DROPPING LEAD
Bullets and lead shot used to be made by letting drops of molten lead fall from the top of a "shot tower". While still liquid, the lead droplets formed spheres, and then "froze" in this shape.

A WEALTH OF SUBJECTS
John Tyndall (1820-1893) was interested in how heat causes changes of state. He also studied a huge range of other subjects, including the origins of life, and why the sky is blue.

1 SOLID STATE
Like most other substances, water can exist as a solid, liquid, or gas. The solid state is ice, and it forms when liquid water is cooled sufficiently. Ice may look different from water, but chemically it is exactly the same.

Pump to remove air from large flask

Stopcocks open to let air flow from small flask to large one

REDUCING THE PRESSURE
Tyndall's apparatus, shown here, demonstrates that water that is not hot enough to boil at ordinary atmospheric pressure will begin to boil when the pressure on it is reduced.

Water is placed inside this flask

Bunsen burner, the heat source

Almost all air is removed from this flask

Each piece of ice has a definite shape

PRESSURE AT WORK
This is a cast-aluminium pressure cooker from about 1930. The very high pressure in a pressure cooker allows water to be heated above its normal boiling point, and the food inside is quickly cooked.

Pressure gauge

Safety valve

Holes formed by bubbles of gas

STONE FROM A VOLCANO
Pumice stone is molten lava which has cooled very quickly. It is honeycombed with holes, which are "frozen-in" bubbles of gas.

MOLTEN MOUNTAIN
When a volcano erupts, it can violently expel thousands of tonnes of lava – red-hot molten rock from the Earth's core. As the lava cools it changes state and solidifies.

HIDDEN HEAT
Joseph Black (1728-1798) measured the heat needed to turn a solid into a liquid, or a liquid into a gas. He called this heat "latent", or hidden.

2 LIQUID STATE
When ice is warmed, it can turn into liquid water. This change happens at a definite temperature, which is normally 0° C (32° F). Under normal pressure water stays liquid up to 100° C (212° F).

3 GAS STATE
When water is heated sufficiently, it starts to turn into steam – a colourless, invisible gas. It can be "seen" only as bubbles in water. What is usually called steam is really a fine mist of water droplets.

Gas turns into liquid water where it touches a cooler surface

A liquid's surface is horizontal

Steam is invisible

Bubbles of steam form in the liquid

A liquid takes the shape of its container

Colloids and glasses

SOME MATTER IS DIFFICULT to classify. For example, lead, a metal, flows like a liquid over centuries. Glass, a seemingly solid substance, is actually a supercool liquid, and flows over decades. The atoms (pp. 34-35) in such substances are not firmly locked into a regular pattern. Instead they form a disorderly pattern, and the atoms move around, allowing the substance to flow. In a form of matter called a colloid, one substance is dispersed through another. The dispersed particles are much larger than atoms, but too small to be seen by the naked eye. Colloids include coloured glass (solid particles dispersed in a solid), clay (solid in liquid), smoke (solid in gas), milk (liquid in liquid), mist (liquid in gas), and foam (gas in liquid).

Carved obsidian makes a sharp arrowhead

NATURAL GLASS
Obsidian forms from molten volcanic rock. The rock cools quickly, and the atoms cannot form a regular pattern. Ancient peoples used obsidian for arrowheads like the one above.

Molten glass

Strong shears

Measuring mould

GLASSBLOWING
Glass is made by melting sand mixed with other ingredients, and then cooling the liquid rapidly. It was first made around 4000 BC in the Middle East. Glass was blown to fit tightly inside a mould from the 1st century AD. Most glassblowing is now done mechanically, but a traditional method is shown in the following sequence. It is now practised only for specialized objects.

Soda ash (sodium carbonate)

Sand (silica)

Limestone (calcium carbonate)

1 A GLASS RECIPE
The main ingredient in the recipe for glass, called the batch, is sand. The next main ingredient is usually soda ash (sodium carbonate), which makes a glass that is fairly easy to melt. Limestone may be used to produce a water-resistant glass.

Iron oxide gives a green colour

Barium carbonate gives a brown colour

2 CUTTING GLASS
A large quantity of the molten glass is gathered on the end of an iron rod by the glassmaker. It is allowed to fall into a measuring mould, and the correct quantity is cut off, using a pair of shears. There are several other traditional glassmaking techniques. Flat glass for windows can be produced by spinning a hot molten blob of glass on the end of a rod. It is spread into a large disc, from which flat pieces can be cut. Glass with a decorated surface can be made by pressing molten glass into a mould.

24

A GOOD TURN
This 18th-century glassmaker is remelting and turning the rim of a glass before correcting its shape. Glass gradually becomes softer over a range of temperatures. In its semi-solid state, glass can easily be shaped into the desired form.

GAFFER AND SERVITOR
The gaffer, or master glassmaker, is making a thin stem for a drinking-glass. He rolls the iron rod on the arms of his chair, to keep the glass symmetrical. The servitor, or assistant, draws out one end of the glass with a rod, while the gaffer shapes and cuts the stem.

Hollow blowing-iron

Shaping mould

Layer of steam cushions the glass

Parison

Flat plate

Final product shows no signs of the joint between the two halves of the mould

3 **MAKING THE PARISON**
The correct quantity of molten glass is picked up from the measuring mould on a hollow blowing-iron and is reheated in a furnace. The glassmaker blows a little air through the blowing-iron and taps the glass on a flat plate a few times to shape it. The glass is now approximately the size and shape of the finished product – in this case a bottle – and is called a parison. Behind the parison is the open shaping mould, which is bottle-shaped. The parison is now ready to be placed inside.

4 **BLOWING, MOULDING, AND SPINNING**
When the mould has been tightly closed, the glass is gently blown again. The glass expands and takes the shape of the inside of the mould. As well as blowing, the glassblower also spins the blowing-iron rapidly. This ensures that the final object does not show any signs of the joint between the two halves of the mould, or any other defects. The glass never comes in direct contact with the material of the mould. This is because the inside of the mould is wet and a layer of steam forms, cushioning the glass.

5 **ONE BROWN BOTTLE**
The shaping mould is opened to reveal the final product – a reproduction 17th-century bottle. This specialized object has to be broken away from the blowing-iron. The jagged mouth of the bottle has to be finished off by reheating it in a furnace and using shaping tools. Since the glass has cooled slightly, the rich brown colour provided by the special ingredients in the batch is revealed. In the very early days of glass, it was always coloured. The first clear glass was made in the 1st century BC.

Mixtures and compounds

Wʜᴇɴ ꜱᴀʟᴛ ᴀɴᴅ ꜱᴀɴᴅ ᴀʀᴇ ᴍɪxᴇᴅ ᴛᴏɢᴇᴛʜᴇʀ, the individual grains of both substances can still be seen. This loose combination of substances is called a mixture. The mixture of salt and sand is easy to separate – if it is given a gentle shake, the heavier grains of sand settle to the bottom. Mixing instant coffee and hot water produces a closer combination, called a solution. Yet this is still fairly easy to separate. If the solution is gently heated, pure water is given off in the form of water vapour, while solid coffee is left behind. The closest combinations of substances are chemical. When carbon (in the form of charcoal) burns, oxygen from the air combines with it to form the gases carbon dioxide and carbon monoxide. These gases are difficult to break down, and are called compounds.

WHEAT FROM CHAFF
Traditionally, wheat was threshed to loosen the edible grains from the chaff (the husks). The grains and chaff formed a mixture that could be separated by "winnowing". Grains were thrown into the air and the breeze blew away the lighter chaff, while the grain fell back.

GOLD IN THE PAN
Gold prospectors in the 19th century would "pan" for gold. They swilled gravel from a stream-bed round a pan with a little water. Any gold nuggets present would separate from the stones because of their high density.

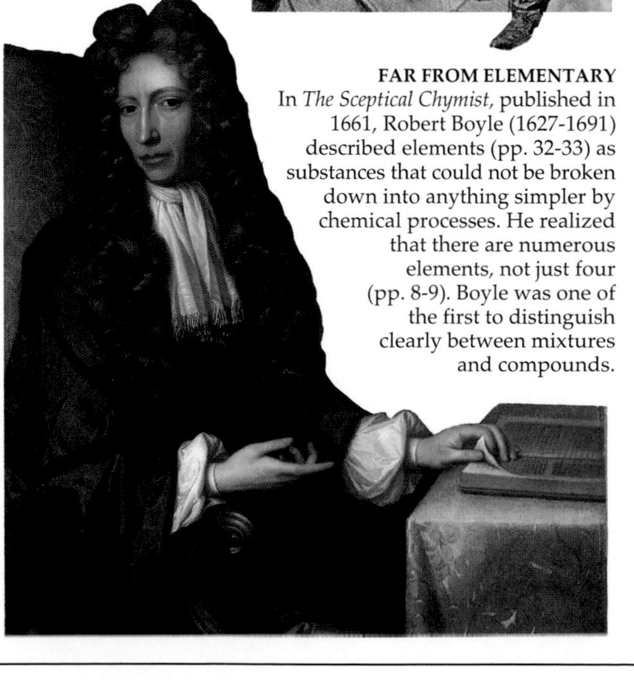

FAR FROM ELEMENTARY
In *The Sceptical Chymist,* published in 1661, Robert Boyle (1627-1691) described elements (pp. 32-33) as substances that could not be broken down into anything simpler by chemical processes. He realized that there are numerous elements, not just four (pp. 8-9). Boyle was one of the first to distinguish clearly between mixtures and compounds.

COLOURFUL REPORT
Mixtures of liquids or gases can be separated by chromatography. The blotting paper shown here has been dipped into an extract of flower petals. Some of the liquid is drawn up into the paper, but the components flow at different rates, and separate out into bands of colours.

Blotting paper

Fastest moving component

Mashed petals and white spirit

ANALYSING COMPOUNDS

This condenser was invented by Justus von Liebig (1803-1873) around 1830, for analysing carbon-containing compounds. The compound was heated to turn it into a gas, and passed over copper oxide in the glass tube. Oxygen from the copper oxide combined with carbon and hydrogen in the gas, and formed carbon dioxide gas and water vapour. The potassium hydroxide in the glass spheres absorbed the carbon dioxide. The amount of carbon in the original compound could be worked out by the increase in weight in the spheres.

REFLECTIVE THINKER

Justus von Liebig made many advances in the chemistry of "organic" substances. This originally meant substances made in living organisms, but now refers to most carbon-containing substances (pp. 40-43). His scientific feats included devising standard procedures for the chemical analysis of organic compounds, inventing a method of making mirrors by depositing a film of silver on glass, pioneering artificial fertilizers, and founding the first modern teaching laboratory in chemistry.

Charcoal is used as the heat source

RUSTING AWAY

The compound rust is a red solid. It forms when iron, a greyish solid, combines with the gases oxygen and hydrogen. When iron is exposed to air, rust forms spontaneously, but the reaction is not easily reversed – the compound can only be broken down again by chemical means.

Sample experiences a strong force

Glass tube contains copper oxide

Potassium hydroxide in the glass spheres absorbs the carbon dioxide

SALT OF THE EARTH

In the 19th century salt was shown to be a compound of two previously unknown substances – sodium, a silvery metal, and chlorine, a poisonous gas.

IN A WHIRL

Mixtures of liquids, or suspensions of solids in liquids, naturally separate out over a period of time. This process can be speeded up by whirling the sample round in a centrifuge.

Handle is connected to the shaft by gears so the speed of rotation is increased

Measuring device

Metal holders for the test tubes

Calcium chloride in this tube absorbs water, allowing the amount of hydrogen in the compound to be calculated

Test tube

SALT OF THE SEA

These salt pans in India are shallow pits that are flooded with sea-water (a mixture of salt and water). The water evaporates in the hot sunshine, but only pure water comes off as a vapour. The salt is left behind as a white solid.

Hand centrifuge in action

Hand centrifuge is operated by turning the handle

Components of the hand centrifuge

Conservation of matter

MATTER COMBINES, separates, and alters in countless ways. During these changes, matter often seems to appear and disappear. Hard deposits of scale build up in a kettle. Water standing in a pot dries up. Plants grow, and their increase in weight is much greater than the weight of the water and food that they absorb. In all everyday circumstances matter is conserved – it is never destroyed or created. The scale found in the kettle built up from dissolved matter that was present in the water all the time. The water in the pot turned into unseen gases that mingled with the air. The increased bulk of the plants came from the invisible carbon dioxide gas in the air. Only in nuclear explosions, or in the Sun and stars, or in other extreme situations, can matter be created or destroyed (pp. 62-63).

GREAT SURVIVOR
The original matter in a long-dead organism is dispersed and survives. A fossil is the last visible trace of an organism.

Ornamental pan scales

THE BALANCE OF LIFE
Lavoisier weighed people and animals over long periods of time to discover what happened to their air, food, and drink. He calculated the quantities of gases involved by examining the measured quantities of solids and liquids that they had consumed.

WEIGHTY MATTER
In the late 18th century the balance became the chemist's most important measuring instrument. Accurate weighing was the key to keeping track of all the matter involved in a reaction. It led to the abandonment of the phlogiston theory (pp. 30-31) – that when a material burns, a substance called phlogiston is always released.

Glass dome traps gases

Fresh pear

WEIGHING THE EVIDENCE
Lavoisier's theory of the conservation of matter can be effectively demonstrated by comparing the weight of substances before and after an experiment. Here, a pear is placed under an airtight container and weighed. The pear is left for a few days and then weighed again. The two weights can be compared to discover whether the process of decay has involved any overall weight change.

A COUPLE OF CHEMISTS
Antoine Lavoisier (1743-1794) stated the principle of conservation of matter in 1789. This was not a new idea – matter had been assumed to be everlasting by many previous thinkers. Lavoisier, however, was the first to demonstrate this principle actively. His wide-ranging investigations were renowned for their rigour – he carried out experiments that were conducted in sealed vessels, and made accurate records of the many substances involved in chemical reactions. This work was extremely careful and laborious, but he was aided by another talented chemist and devoted co-worker, his wife, Marie-Anne.

Pointer indicates that the pans are perfectly balanced

Potassium permanganate dissolves and forms a solution

Water

NATURAL EROSION
The land is constantly worn away by wind, rain, and waves, yet this is balanced by the natural building up of new land forms elsewhere. No matter is lost or gained overall.

Potassium permanganate crystals

AN OBVIOUS SOLUTION
Solids left in water often dissolve. If the solids are colourless, such as salt, it is easy to believe they have disappeared completely. In fact they have just thoroughly mixed, and broken into minute particles, which have spread through the liquid. When the solid is coloured, like this potassium permanganate, it is easier to believe that it still exists in the liquid. Weighing the solution confirms that it weighs the same as the original liquid and solid.

OUT WITH A BANG
When a firework goes off, gunpowder burns, as well as other chemicals, plus the cardboard and paper of the firework packaging. The burning products form gases, and a small quantity of solids. Though widely scattered, the combined products weigh the same as the original firework.

Glass dome contains air and gases produced by rotting pear

Condensation

ROTTEN RESULT
After a few days, rotting begins to take place, and some parts of the pear become brown and mushy. In the air under the glass dome there is now less oxygen, for some of it combines with the substances in the pear. There is more carbon dioxide, however, as well as other gases released by the fruit. Overall, the weight of the container plus its contents does not alter in the slightest degree. Early chemists did not realize that if the glass dome is lifted before weighing, air is likely to enter or escape, and therefore affect the weight of the container and its contents.

Rotting pear

Scale pan

Burning matter

ONE OF THE FIRST great achievements of 18th-century science was the explanation of burning (also known as combustion). Georg Stahl (1660-1734) put forward the theory that an element, phlogiston, was given out in burning. His theory was wrong – it would mean that all substances would lose weight when they burn. Several chemists had already observed that some substances such as metals increase in weight during burning, and the theory of phlogiston was firmly denied by Antoine Lavoisier (pp. 28-29). He argued that air contains a gas that combines with a substance when it burns, and named the gas oxygen. Sometimes substances can "burn" in gases other than oxygen. Some, such as ammonium dichromate, can change by themselves into other substances, producing flame, heat, and light.

(pp. 28-29)

FROM CRYSTALS TO ASH
In the following sequence, orange ammonium dichromate crystals produce flame, heat, and light, and turn into grey-green ash.

Low flames

Orange crystals of ammonium dichromate

Ash quickly forms

1 READY TO REACT Ammonium dichromate is a substance used in indoor fireworks. It consists of nitrogen, hydrogen, chromium, and oxygen.

2 VITAL SPARK When lit by a flame, the substance's atoms (pp. 34-35) form simpler substances, and produce heat and light.

(pp. 34-35)

Brass pivot

Burning-lens is angled to catch sunlight

Wooden stand

Melting ice in flask

THE HEAT OF THE MOMENT
Antoine Lavoisier was particularly interested in chemical reactions that required great heat. One problem in his scientific work was to obtain heat that was both intense and "clean", for often the reacting substances were contaminated by smoke and soot from the heat source (usually a flame). His solution was this giant mobile burning-glass, or convex lens, with which he enthralled the French populace in 1774.

FOCUS OF ACTIVITY
Heat brings about many changes in matter. It can cause different substances to react together, or it can make a reaction go faster. Here the heat is produced as sunlight is focused by a large convex lens to fall on to a flask containing ice. This causes the ice to melt – a physical rather than a chemical change. If sunlight is focused on to paper, the paper can smoulder, and even burst into flame. This is a chemical change, and is an example of combustion.

3 BREAKDOWN
The substance is rapidly converted into chromium oxide, a compound of chromium and oxygen, and into nitrogen and water vapour, both invisible gases.

Higher, stronger flames

4 ASHEN ENDING
The orange crystals of ammonium dichromate have broken down, leaving a large pile of chromium oxide. The nitrogen and water vapour have escaped into the air.

Grey-green ash of chromium oxide

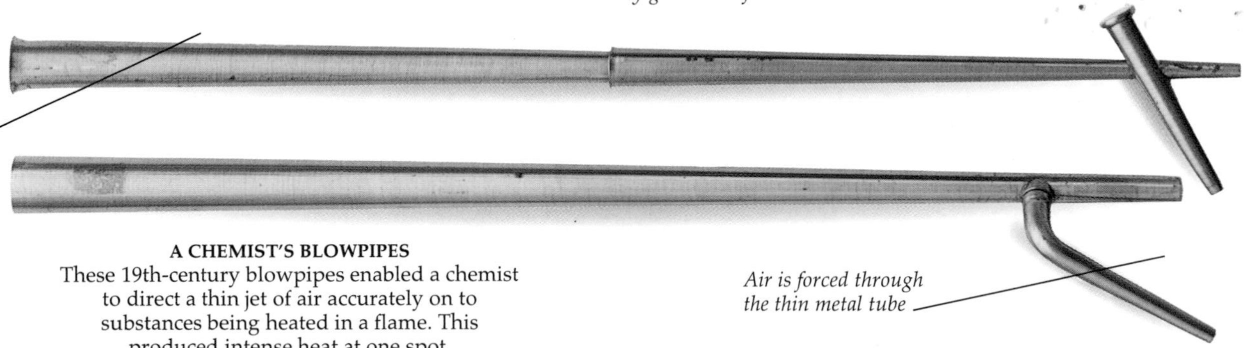

Air is blown through the mouthpiece

A CHEMIST'S BLOWPIPES
These 19th-century blowpipes enabled a chemist to direct a thin jet of air accurately on to substances being heated in a flame. This produced intense heat at one spot.

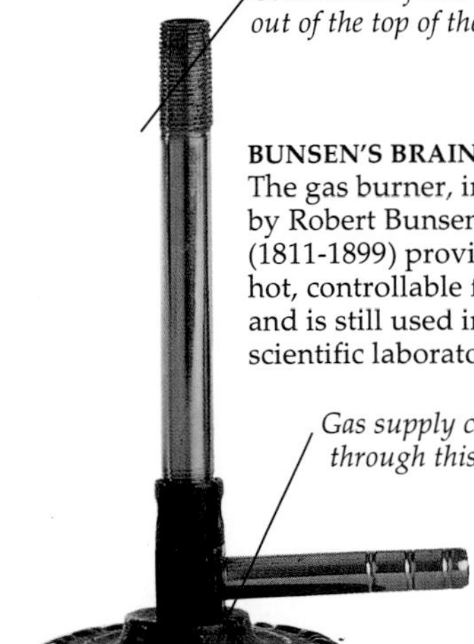

Air is forced through the thin metal tube

Controllable flame comes out of the top of the burner

Large surface area increases the amount of heat that can be delivered

BUNSEN'S BRAINCHILD
The gas burner, invented by Robert Bunsen (1811-1899) provides a hot, controllable flame, and is still used in scientific laboratories.

Gas supply comes through this pipe

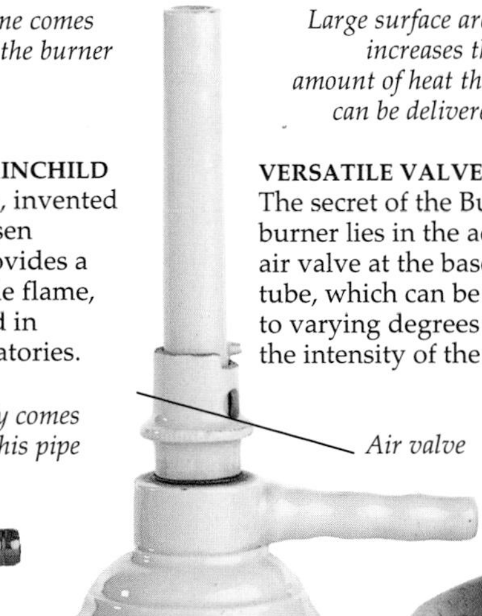

VERSATILE VALVE
The secret of the Bunsen burner lies in the adjustable air valve at the base of the tube, which can be opened to varying degrees to alter the intensity of the flame.

Air valve

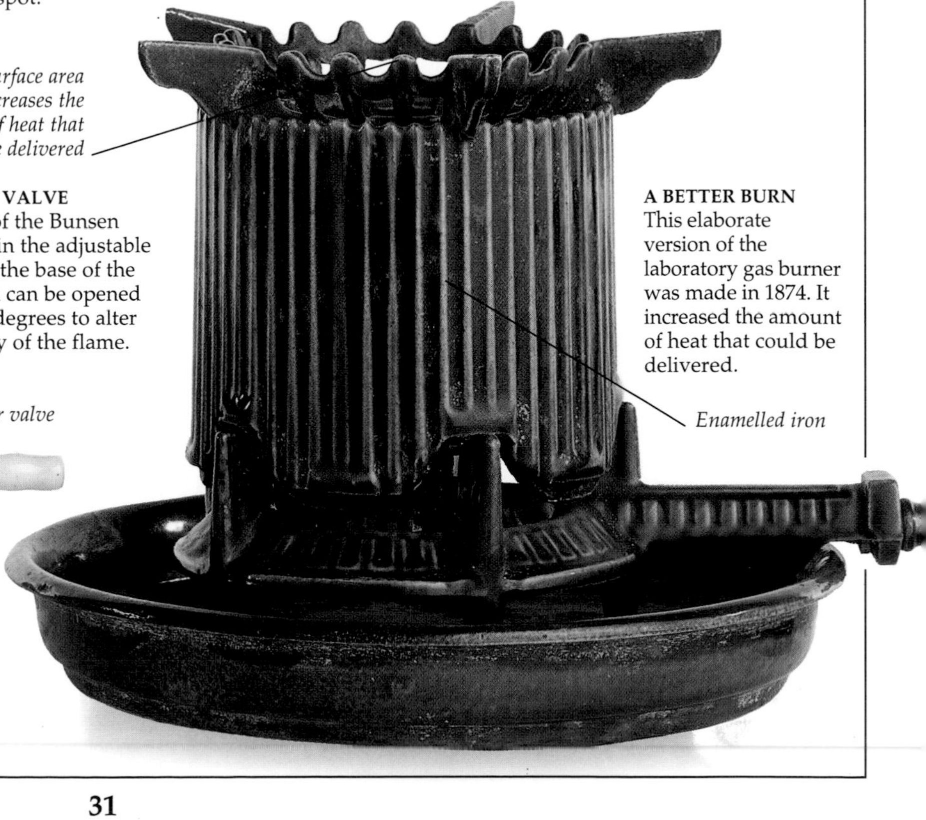

A BETTER BURN
This elaborate version of the laboratory gas burner was made in 1874. It increased the amount of heat that could be delivered.

Enamelled iron

Bunsen burner from 1889

Bunsen burner made from flame resistant porcelain

Charting the elements

Eʟᴇᴍᴇɴᴛs ᴀʀᴇ ᴘᴜʀᴇ sᴜʙsᴛᴀɴᴄᴇs – they do not contain anything else, and cannot be broken down into simpler substances. Many of the elements were discovered during the 18th and 19th centuries, particularly by using processes such as electrolysis and spectroscopy. In electrolysis, an electric current is passed through compounds to break them down (pp. 50-51). In spectroscopy, the light given out by hot substances is analysed with a spectroscope (pp. 56-57) to show an element's characteristic pattern of colours. Dmitri Mendeleyev (1834-1907) brought order to the elements with his "periodic table", based on patterns in properties of elements such as their reactivity.

FLAME TEST
Here common salt (sodium chloride) burns yellow in a flame, revealing the presence of the element sodium.

A BATTERY OF DISCOVERIES
After learning of Alessandro Volta's invention of the electric battery in 1800, Humphry Davy (1778-1829) built his own. It was large, and had 250 metal plates. He used it for electrolysis, and prepared pure samples of new metals.

SPLITTING UP SALT
Humphry Davy discovered sodium by electrolysing molten salt in this apparatus. Davy used electrolysis to obtain other metals with similar properties to sodium, such as barium, potassium, magnesium, calcium, and strontium. He then used potassium to extract another new element, boron.

BURNING QUESTION
This box of flame test equipment dates from the 19th century. It includes a blowpipe, tweezers, and different chemicals for testing. In a flame test, tiny quantities of a substance are held on a wire, which is put in to a flame. The colour of the flame often indicates the substance's identity. For example, a flame is turned violet by potassium, and blue-green by copper. Flame tests require the hot flame of the Bunsen burner (p. 31).

Terminal linked to battery

ELECTRICAL BREAKDOWN

Common salt consists of sodium and chloride ions – positively charged sodium atoms, and negatively charged chlorine atoms (pp. 50-51). When salt is melted, the ions move around each other. If metal plates connected to a battery are placed in the molten salt, the positive plate attracts the chloride ions, and the negative plate attracts the sodium ions.

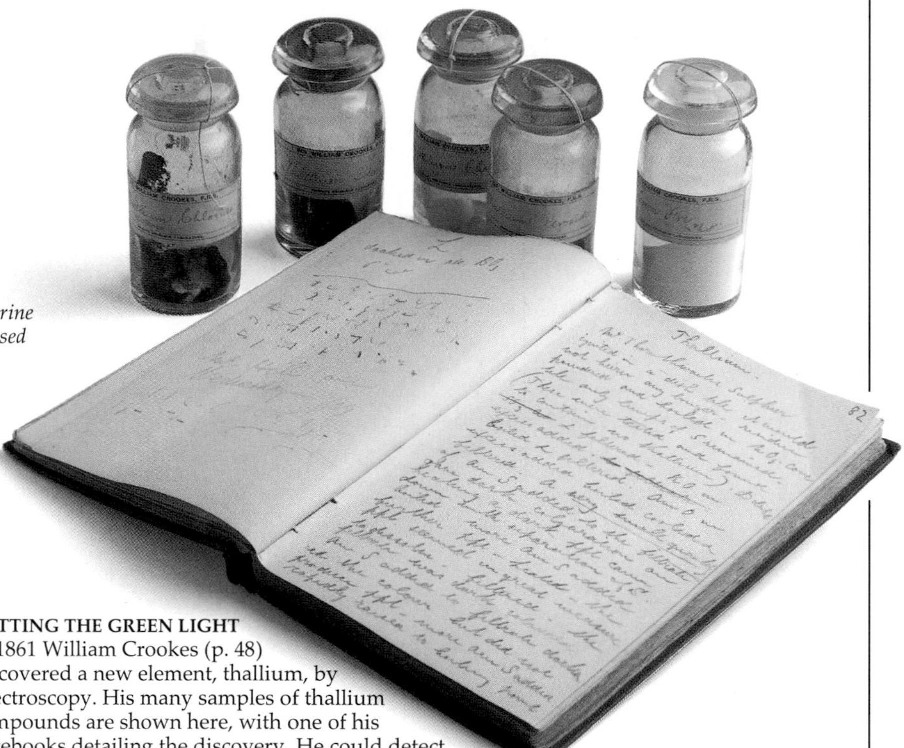

Sodium deposited

Negatively charged chloride ion

Chlorine released

Positively charged sodium ion

Negative terminal linked to battery

Positive terminal linked to battery

Liquid compound was placed in glass dish to be electrolysed

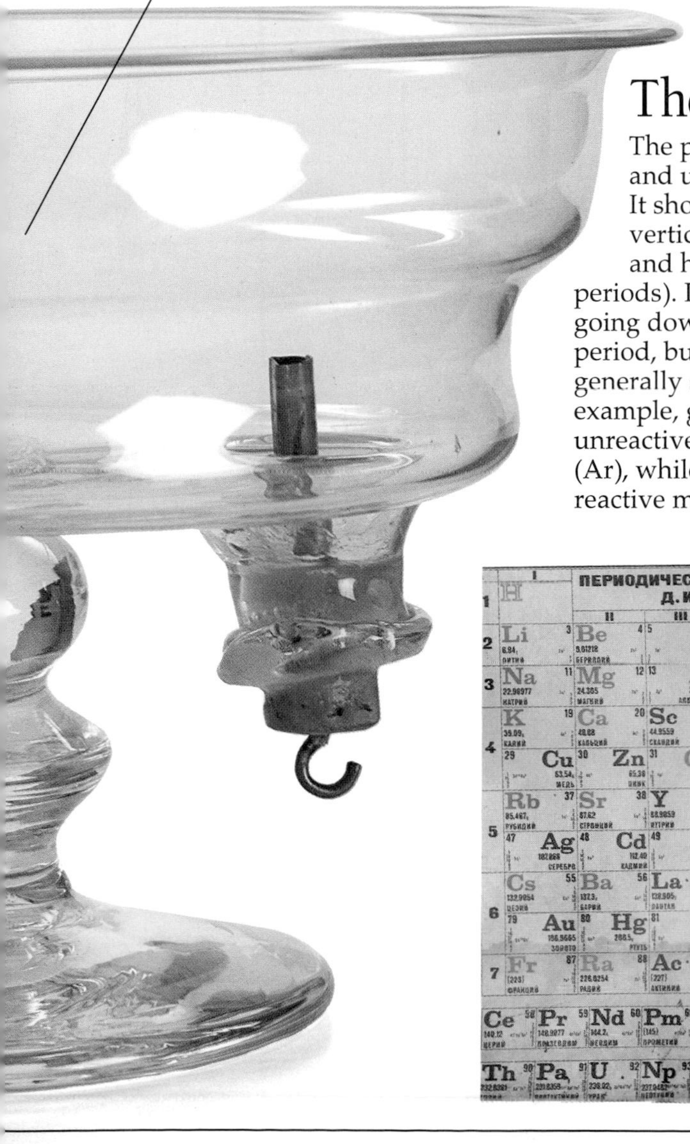

GETTING THE GREEN LIGHT
In 1861 William Crookes (p. 48) discovered a new element, thallium, by spectroscopy. His many samples of thallium compounds are shown here, with one of his notebooks detailing the discovery. He could detect minute quantities of the new element because it emits a brilliant green light when in a hot flame.

The periodic table

The properties of the elements can be described and understood in terms of the periodic table. It shows more than 100 elements, arranged vertically into columns (called groups) and horizontally into rows (called periods). Properties change systematically going down each group and along each period, but elements in each group have generally similar properties. For example, group VIII contains the very unreactive "noble" gases such as argon (Ar), while group I contains very reactive metals such as sodium (Na).

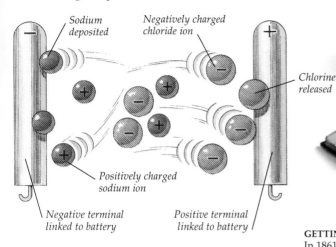

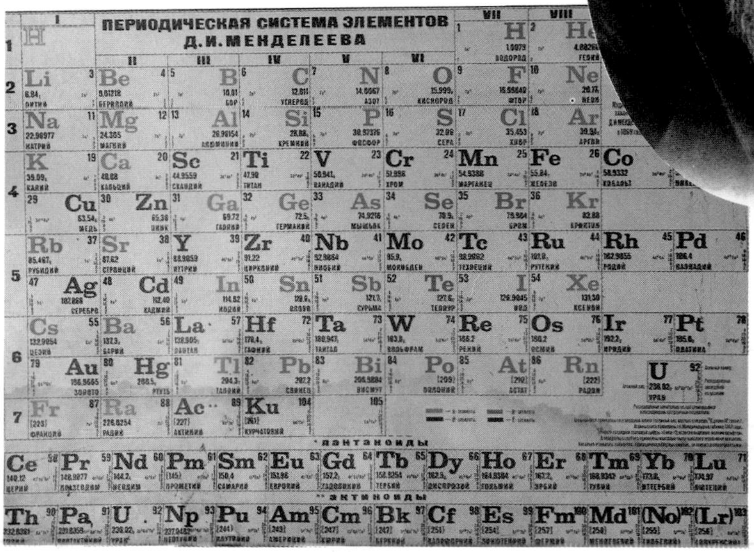

PATTERN FINDER
Dmitri Mendeleyev's periodic table suggested corrections to previously accepted chemical data, and successfully predicted the existence of new elements. The Russian periodic table shown here is based on Mendeleyev's original of 1869.

The building blocks

As SCIENTISTS DISCOVERED more elements, they pondered over the ultimate nature of matter. The ancient idea of atoms (pp. 8-9) received a powerful boost from John Dalton (1766-1844) in 1808. He proposed that each element has its own unique atom, and that each chemical compound is formed by a certain combination of atoms. He showed that the weights of atoms relative to each other could be found by weighing the elements that combined in particular compounds. The comparative weight of an atom could be found, but not the actual weight – an atom could only be said to be so many times heavier than, for example, hydrogen, the lightest atom.

CLOSING IN ON ATOMS
Atoms make up the world in much the same way that the letters of the alphabet make up a book. Scientists need a close-up view to study atoms, just as readers have to peer closely at a page to study individual letters.

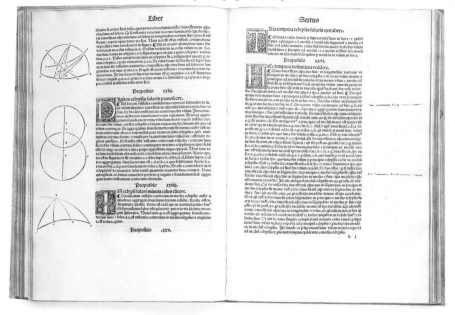

1 THE BOOK OF THE WORLD
A glance at any book shows that it is made up of many things, such as pictures, text of large and small print, and different chapters. Similarly, a glance at the "book of the world" shows that it is a kaleidoscopic array of many sorts of chemical substances. But such a glance does not reveal whether or not the world is made of atoms.

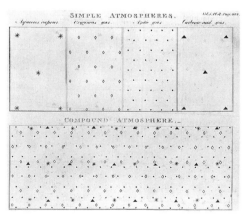

ATMOSPHERIC ATOMS
John Dalton drew these diagrams in 1802. He was a keen meteorologist and knew that air consists of several gases – oxygen, water vapour, carbon dioxide, and nitrogen (top diagram). He developed his atomic theory while explaining why these gases remain mixed, rather than forming separate layers.

2 A PAGE AT A TIME
If a reader concentrates on one page of the "book of the world" and temporarily ignores the rest of the pages, this still gives a sample of the world that is very large compared with the atom. Similarly, when studying matter scientifically a small part must be isolated, for example, by examining a substance in a flask in a laboratory. To obtain a more detailed view of matter, scientists need to use instruments.

ELASTIC FLUIDS
Dalton assumed that gases are made of atoms which are far apart and can move independently. This is why gases can be compressed and expanded. He called gases "elastic fluids", and pictured their atoms like these circles.

John Dalton

Drawn & Etched by J. Stephenson

3 ENDLESS VARIETY
In close-up, one piece of text consists of many different words. Similarly, with the aid of chemical analysis and instruments, matter can be seen to be made up of an enormous number of different substances.

4 NARROW VIEW
A microscope can give a detailed view of a small sample of matter, but this sample may be composed of a variety of substances. It is similar to a sentence, which is made up of many different words.

DALTON'S ATOMS AND ELEMENTS
In 1808 John Dalton published his atomic theory. It suggested that all matter is made up of indivisible atoms; each element is composed of atoms of characteristic weight; and compounds are formed when atoms of various elements combine in definite proportions. Below are Dalton's symbols for the atoms of the 36 elements that he believed to exist (over 100 elements have now been found). Some of Dalton's elements shown here, for example lime and soda, are actually compounds, not elements. Dalton also calculated the weight of each element's atom, by comparing it to hydrogen.

Ratio

5 WORDS OF SUBSTANCE
"Words" in the "book of the world" are groups of atoms, or molecules (pp. 36-37). Here, the 26 letters of the Roman alphabet make up words. About 90 different kinds of atom make up molecules.

ELEMENTS

	Wt.			Wt.
Hydrogen	1	Strontian		46
Azote	5	Barytes		68
Carbon	54	Iron		50
Oxygen	7	Zinc		56
Phosphorus	9	Copper		56
Sulphur	13	Lead		90
Magnesia	20	Silver		190
Lime	24	Gold		190
Soda	28	Platina		190
Potash	42	Mercury		167

Dalton's elements

Dalton's carbon atom

6 SINGULAR CHARACTERS
The letters on the printed page correspond to atoms. Just as letters group into words, so atoms form molecules. There is no limit to the words that can be formed from the alphabet, and any number of compounds can be formed from atoms. Not all possible combinations of letters are permitted, and neither are all combinations of atoms.

Molecules

ATOMS CAN EXIST SINGLY in some gases, but in many substances they form groups called "molecules". For example, the molecule of water consists of an oxygen atom (O) joined to two hydrogen atoms (H). Its chemical formula is H_2O. Some molecules can be much bigger than this, containing thousands of atoms. In the mid 19th century it was realized that chemical bonds could explain the ways in which atoms link together to form molecules. A bond is like a hook that can link to a similar hook on another atom. For example, an atom of the gas nitrogen has three hooks, while a hydrogen atom has one. Each bond on the nitrogen atom can link to the bond on one hydrogen atom, which produces the molecule NH_3, the gas given off by smelling salts.

19th-century molecular model of monochloromethane, a type of solvent

Ammonia molecule (NH_3)

Chemical bond

Oxygen

Sulphur

Nitrogen

Hydrogen

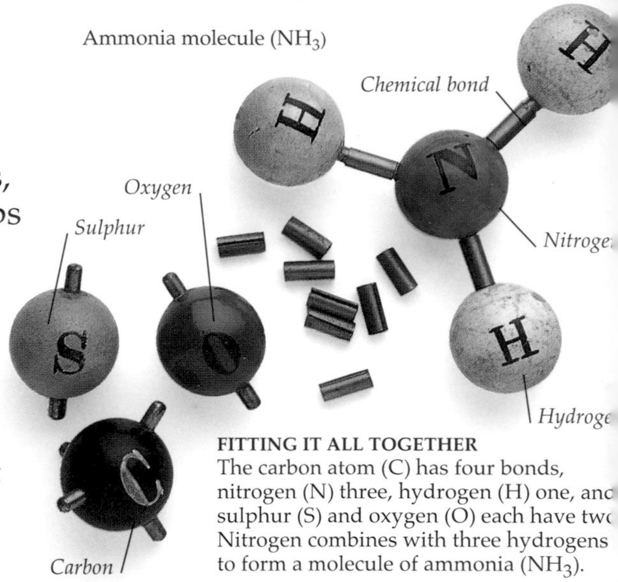

FITTING IT ALL TOGETHER
The carbon atom (C) has four bonds, nitrogen (N) three, hydrogen (H) one, and sulphur (S) and oxygen (O) each have two. Nitrogen combines with three hydrogens to form a molecule of ammonia (NH_3).

Carbon

Calcium carbonate molecule ($CaCO_3$)

Carbon

Oxygen

Calcium

1 THE RAW MATERIAL
Limestone is a whitish rock with the chemical name of calcium carbonate. As this name suggests, the molecule of limestone contains atoms of calcium and carbon, but it also contains oxygen. Each atom of carbon is tightly joined to three of oxygen, and this group is more loosely linked to one atom of calcium.

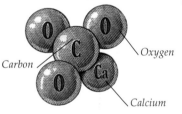

Limestone (calcium carbonate)

Limestone breaks up when heated

Bunsen burner

Pestle *Mortar* *Calcium hydroxide paste*

ITALIAN IDEAS

Amedeo Avogadro (1776-1856) suggested that in equal volumes of any two gases there are always the same number of molecules, if the gases are at the same temperature and pressure. For about 50 years his idea was largely ignored, until another Italian chemist, Stanislao Cannizzaro (1826-1910), publicized it. The idea was then quickly accepted, and it helped to clarify many chemical reactions.

ORCEFUL SUGGESTION

ns Jakob Berzelius 779-1848) was one of the rst people to suggest that oms are held together in olecules by electrical rces (pp. 60-61).

Crumbly powder (calcium oxide)

Water

Pipette

Hydrogen

Calcium

Oxygen

Calcium hydroxide molecule (Ca(OH)$_2$)

Calcium carbonate molecule (CaCO$_3$)

Oxygen

Carbon

Calcium

3 ADDING WATER

When water is added to calcium oxide powder there is a strong reaction, and the powder swells up and gives out heat. The molecules of the calcium oxide and of the water, H$_2$O, rearrange themselves to form molecules of calcium hydroxide, a soft and pasty substance. As its name suggests, this molecule contains calcium, hydrogen, and oxygen. Its formula is Ca(OH)$_2$, showing that two oxygen-hydrogen pairs (OH) are joined to a calcium atom.

4 BACK TO CALCIUM CARBONATE

The calcium hydroxide dries and hardens. Water molecules (H$_2$O) are given off into the air, while carbon dioxide molecules (CO$_2$) from the air are absorbed. The calcium hydroxide (Ca(OH)$_2$) turns into calcium carbonate (CaCO$_3$), chemically identical to the raw material. The reconstituted calcium carbonate looks different from the original limestone because it was not formed under high pressure within the Earth.

arbon dioxide olecule (CO$_2$)

arbon

Oxygen

Calcium

Calcium oxide molecule (CaO)

HEATING THE LIMESTONE

When limestone is heated, it turns into a ft, crumbly powder called calcium oxide. This ppens because each molecule of the original lcium carbonate breaks into two smaller olecules. One of these molecules nsists of the calcium atom (Ca) joined a single oxygen atom (O), making aO. The other molecule consists of e carbon atom joined to the other two xygen atoms, making CO$_2$. This is rbon dioxide gas, which escapes into e atmosphere.

Tongs

Reconstituted calcium carbonate

Molecules in motion

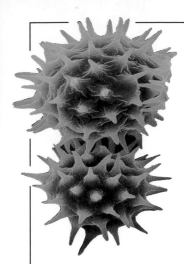

Highly magnified pollen grains – the key to molecular movement

U P UNTIL THE MID 18TH CENTURY, there was a widely accepted theory that heat was a "fluid" called caloric. However, in 1799 Count Rumford (1753-1814) observed that limitless quantities of heat could be generated in the boring of cannon barrels. He suggested that the drilling work was increasing the motions of the atoms that made up the metal. This idea gained support when James Joule (1818-1889) did experiments to measure exactly how much work was needed to generate a definite amount of heat. When heat is applied to matter, the motion of the molecules is increased, and the temperature rises. Gradually it was realised that the differences between the three states of matter – solids, liquids, and gases (pp. 22-23) – are caused by the motion of molecules. The molecules in a solid are fixed, but can vibrate. The molecules in a liquid move about, but still remain in contact with each other. In a gas the molecules fly about freely, and move in straight lines until they collide with each other or with other objects.

DANCE OF THE POLLEN
In 1827, Robert Brown (1773-1858) observed pollen grains under a microscope. The grains were suspended in a liquid, and were in constant motion. He thought the motion originated in the pollen particles. But Albert Einstein (p. 55) in 1905, and Jean Perrin (1870-1942) in 1909, explained that the grains were being buffeted by the movement of the liquid's molecules.

Pointer moves round the dial to show expansion of the rod

MEASURING HEAT EXPANSION
When a solid is heated, the vibration of its molecules increases. Each molecule then needs more space to vibrate, and the solid expands. This device, a pyrometer (meaning "heat-measurer") from the mid 19th century, showed how a metal rod increased in length as it was heated by a gas flame placed beneath it, and then shrank again as it cooled.

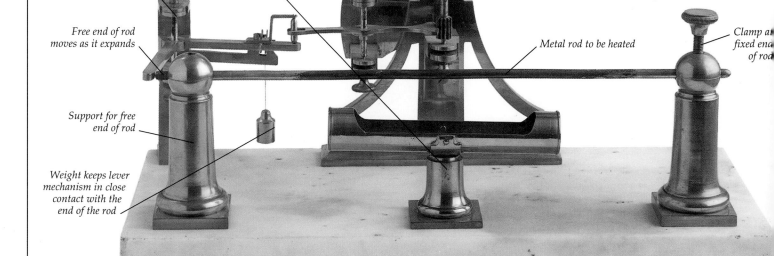

Lever turns when rod changes length

Spirit burner for heating the rod

Free end of rod moves as it expands

Metal rod to be heated

Clamp a fixed end of rod

Support for free end of rod

Weight keeps lever mechanism in close contact with the end of the rod

TURNING TO HEAT
In the 1840s James Joule used this water friction apparatus to measure how much heat a given amount of mechanical "work" could be converted into. The work was done by a weight that turned paddles in a container of water. The fixed vanes limited the swirling of the water, so the work done was converted into heat. Joule measured the water's rise in temperature, and calculated the heat generated. His results added evidence to the theory that heat is the movement of molecules.

Handle to wind up weight

String connected to falling weight turns rod

Paddles

Water inlet

Water outlet

Container insulates apparatus from outside heat

Fixed vanes resist movement of water

TAKING TEMPERATURES
James Joule measured the "rate of exchange" between heat, mechanical work, and electrical energy.

RACING AHEAD
Ludwig Boltzmann (1844-1906) was one of the first scientists to assume that molecules in gases move at a range of speeds (previous scientists had assumed for simplicity that all molecules moved at the same speed). He worked out that gas molecules can rotate and vibrate, as well as move through space.

Glass separating plate

Bromine is brown

UPWARDLY MOBILE
Gases expand to fill available space. Here, the gas bromine, which is heavier than air, is confined in the lower jar. But when the separating plate is removed, the bromine molecules diffuse into the upper jar.

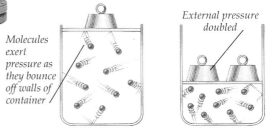

Molecules exert pressure as they bounce off walls of container

External pressure doubled

BOYLE'S LAW
Boyle (p. 26) saw that when a gas is pushed into a smaller volume, it exerts greater pressure. This is because the molecules hit the container walls more frequently.

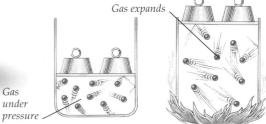

Gas expands

Gas under pressure

CHARLES'S LAW
Charles (p. 21) saw that when a gas is heated, it exerts greater pressure, and will expand if it can. The molecules move faster and collide more violently with the container walls.

Carbon rings and chains

CLOSING THE CIRCLE
The structure of the benzene molecule, a form of carbon, resembles a snake swallowing its own tail.

CARBON IS UNIQUE in the number and complexity of compounds it can form. More than 7 million carbon-containing compounds are now known, compared with about 100,000 compounds made from all the other elements. Carbon is essential to the chemistry of all living things (pp. 42-43). The carbon atom can easily link up with other carbon atoms and with most other types of atom, using its four chemical "hooks", or valency bonds (pp. 36-37). Its molecules can have a "backbone" of a long chain of carbon atoms, either straight or branched. The carbon atoms can also form rings, which can be linked to other rings or carbon chains, to form intricate structures, sometimes consisting of thousands of atoms.

PREHISTORIC PRESSURE
Coal is the fossilized remains of trees and other plants that were buried in swamps. Over about 345 million years, they have been turned into a soft black rock by intense, sustained pressure from layers of other rocks. Coal consists mostly of carbon, with some hydrogen, oxygen, nitrogen, and sulphur. The carbon in coal takes oxygen from the air and burns vigorously, so is a useful source of fuel.

Benzene

SMUDGY CARBON
Charcoal, a form of carbon, is obtained when substances such as wood, bone, or sugar are heated strongly with no air present. Charcoal is soft and easy to smudge, and is an excellent drawing material.

Graphite

PLUMBING THE DEPTHS
Graphite, also known as plumbago, is a form of carbon found as a soft mineral. It can easily be made to split and flake. Graphite is the main component of pencil "leads", and is widely used as a lubricant.

LORD OF THE RINGS
Friedrich Kekulé (1829-1896) tried for a long time to work out how a benzene molecule's six carbon atoms link to the six hydrogen atoms. He found the solution while dozing. He dreamt of a row of carbon and hydrogen atoms closing in a ring, like a snake swallowing its tail.

Diamond molecule

Carbon atoms are arranged in an intricate lattice

Single bond

Diamonds

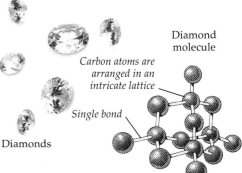

Benzene molecule

Carbon atoms form a ring

Hydrogen atom

Single bond

Double bond

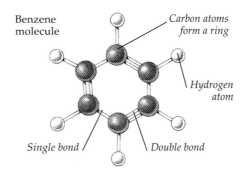

DIAMONDS ARE FOR EVER
The hardest natural material is diamond. It is a valuable gemstone, but is also used as the cutting tip in drills, or for grinding material. It consists of virtually pure carbon. Its atoms are arranged in a very strong three-dimensional lattice (a repeated pattern within a crystal). Each atom is joined to four neighbours by single chemical bonds. Diamonds form where carbon has been subjected to huge geological pressures and temperatures.

BAFFLING BENZENE
When coal is strongly heated, a colourless liquid – benzene – is obtained. The benzene structure is the basis of a huge number of important carbon compounds. Its molecular structure had baffled chemists until Kekulé thought of it as a ring of carbon atoms linked to hydrogen atoms.

FAT FACTS
Butter is a mixture of fats –
carbon-containing substances
that are important in living
things for storing energy.
Similar substances that
are liquid at room
temperature are
called oils.

*Butter is
an edible fat*

Jet brooch

CLEANSING WITH CARBON
Soap is made from substances with
long chains of carbon atoms, usually
15 or 17. One end of each soap
molecule attaches to water and the
other end attaches to oil. This enables
soap to break up oil and grease into
small drops in water.

Petrol

BLACK BEAUTY
Jet is a form of coal called
lignite, and is largely made
up of carbon. It has a deep
velvety black colour, and is
easy to carve and polish. Jet
has been used for jewellery
since early times.

*Each cluster
contains 60
carbon atoms*

Octane
molecule

*Single
bond*

*Carbon
atom*

Graphite
molecule

*Layer of
carbon atoms*

*Hydrogen
atom*

*Double
bond*

*Single
bond*

ATOMS IN LAYERS
Graphite and diamond are the two
crystalline forms of carbon. Unlike
diamond, the carbon atoms in
graphite are joined in flat layers.
Each layer is only weakly joined to
the next, and so the layered atoms
can easily slip over each other.

GLOBULAR CARBON
One of the strangest forms of carbon to be
discovered consists of globular clusters of
carbon atoms, a billionth of a metre across.
The false-colour image shown here is the
simplest form, with each globe containing 60
atoms. It is named buckminsterfullerene,
after Buckminster Fuller (1895-1983), an
architect who developed the similarly
shaped geodesic dome. Though only recently
discovered, buckminsterfullerene (also
known as buckyballs) is fairly common, and
can be found in soot particles.

LIQUID POWER
Most petrol molecules are chains, containing
between five and ten carbon atoms. The petrol
molecule octane (left), contains eight carbon
atoms. Petrol is derived from petroleum oil, a
mixture of liquids, solids, and gases that are
the fossilized remains of microscopic life.

Living matter

Original templates from Watson and Crick's DNA model

ANIMALS AND PLANTS ARE AMAZINGLY complex forms of living matter. They can grow, reproduce, move, and respond to their environments. Until the late 19th century, many scientists thought a "vital principle" must control the behaviour of living matter. Such beliefs changed when scientists began to be able to synthesize a range of "organic" substances (substances previously found only in living things), and started to explain the chemistry of the processes within living creatures. It was once thought that flies and other small creatures could develop spontaneously from rotting matter, but Louis Pasteur (1822-1895) showed that new life can only arise from existing organisms. Life has never yet been made in a scientific laboratory from non-living matter. This leaves the problem of how life developed on Earth from a "soup" of non-living molecules.

FLIES IN THE SOUP
In the 1860s, when Louis Pasteur boiled a flask of broth and left it, he learned that new life arises only from living matter. Dust from the air fell in the broth, and micro-organisms grew. But when only dust-free air reached the broth, nothing grew.

ANALYSING THE ORGANIC
In the 19th century many "organic" substances were closely examined. This apparatus (left) was used in the 1880s to measure the nitrogen in urea.

UNDERSTANDING UREA
A landmark in understanding life was the synthesis of urea, a nitrogen-containing chemical found in animal waste. Friedrich Wöhler (1800-1882) made urea in 1828 from ammonia and cyanic acid.

Saturated solution of common salt was placed in glass cup

Measuring tube was filled with distilled water

Central tube

Sodium hypobromite solution was placed in here

When the tap was opened, the solutions mixed, and nitrogen was produced and collected in the measuring tube

Sample of urea containing unknown concentration of nitrogen was placed in here

Animals take in carbon from green plants, and animal wastes and remains contain carbon

Green plants absorb and give out CO_2

Rocks absorb and give out CO_2

NEVER ENDING CYCLE
Carbon is the basis for all living matter, circulating between air, oceans, rocks, and living things. Carbon dioxide gas (CO_2) is absorbed from the air by green plants. Plant-eaters use the plant's carbon for tissue-building. Carbon is returned to the environment in animals' waste products, and when dead animals decay. Rocks and water also absorb and give off CO_2, and complete the cycle.

Rods represent
chemical bonds

Meeting points of rods
represent atoms

Aluminium plates
represent the four
different bases

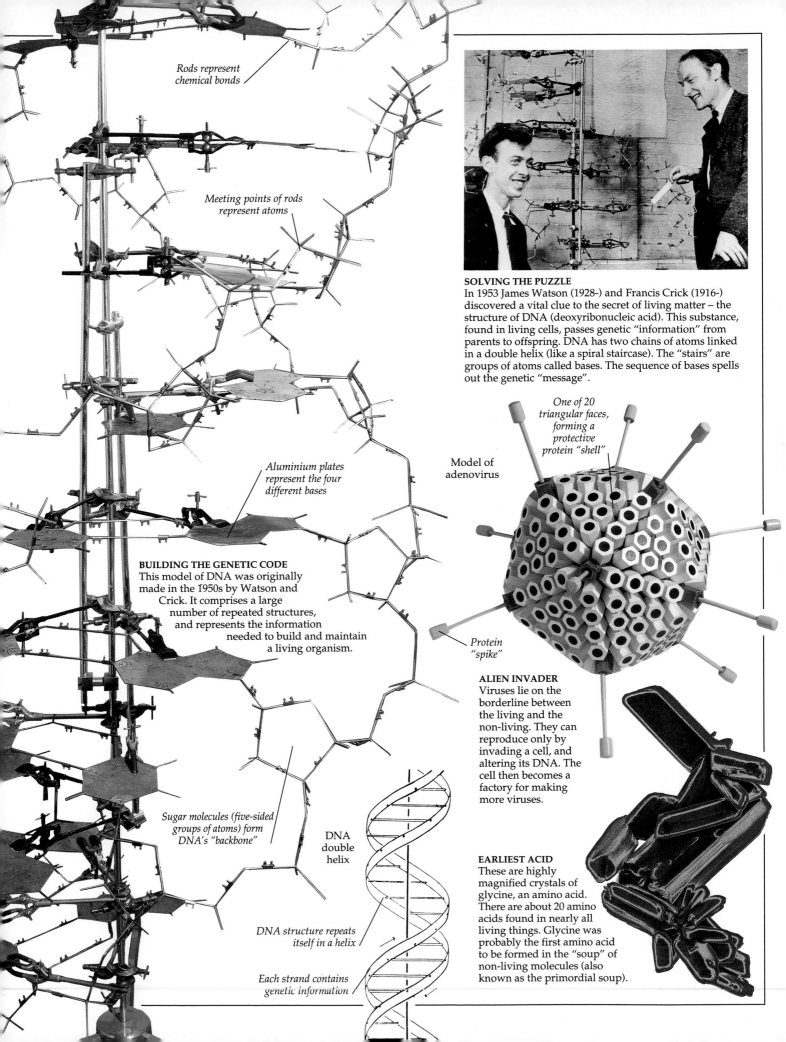

BUILDING THE GENETIC CODE
This model of DNA was originally
made in the 1950s by Watson and
Crick. It comprises a large
number of repeated structures,
and represents the information
needed to build and maintain
a living organism.

Sugar molecules (five-sided
groups of atoms) form
DNA's "backbone"

DNA
double
helix

*DNA structure repeats
itself in a helix*

*Each strand contains
genetic information*

SOLVING THE PUZZLE
In 1953 James Watson (1928-) and Francis Crick (1916-)
discovered a vital clue to the secret of living matter – the
structure of DNA (deoxyribonucleic acid). This substance,
found in living cells, passes genetic "information" from
parents to offspring. DNA has two chains of atoms linked
in a double helix (like a spiral staircase). The "stairs" are
groups of atoms called bases. The sequence of bases spells
out the genetic "message".

*One of 20
triangular faces,
forming a
protective
protein "shell"*

Model of
adenovirus

*Protein
"spike"*

ALIEN INVADER
Viruses lie on the
borderline between
the living and the
non-living. They can
reproduce only by
invading a cell, and
altering its DNA. The
cell then becomes a
factory for making
more viruses.

EARLIEST ACID
These are highly
magnified crystals of
glycine, an amino acid.
There are about 20 amino
acids found in nearly all
living things. Glycine was
probably the first amino acid
to be formed in the "soup" of
non-living molecules (also
known as the primordial soup).

Designing molecules

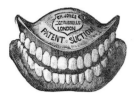

Artificial teeth and dental plate from the 1870s

IN THE MID 19TH CENTURY chemists began to use their new knowledge of organic molecules (pp. 42-43) to make new materials with valuable properties. Parkesine, an imitation ivory, was made in 1862 by Alexander Parkes (1813-1890). In 1884 Hilaire de Chardonnet (1839-1924) made rayon, the first artificial fibre, by imitating the chemical structure of silk. Rubber was toughened and made more useful by the vulcanizing process (heat treatment with sulphur), invented in 1839. The age of plastics was ushered in by Leo Baekeland (1863-1944), who invented Bakelite in 1909. Plastics are polymers – large molecules with possibly thousands of identical groups of atoms chained together. Plastics can be moulded by heat and pressure, but then become fixed in shape. They are unreactive, and do not disturb the body's chemistry when used as a replacement hip-joint, for example. Plastics pose a waste disposal problem, however, for most do not rot. In addition to plastics, modern chemists have designed many other useful products, such as drugs, detergents, and alloys.

Matt bakelite

BAEKELAND'S BAKELITE
The first entirely synthetic plastic, bakelite, was used in early telephones and other electrical devices. It was developed in the United States in 1909.

EBONITE ARTEFACT
This saxophone mouthpiece is made of ebonite, a type of vulcanized rubber. It is also known as vulcanite.

Ebonite resembles ebony

Alluring...Enduring...

Bakelite hair-dryer

RUBBER WEAR
The synthetic rubber isoprene was made in 1892. It is much more resistant to wear than natural rubber. It became very important to the Allies during the Second World War, when rubber plantations in South-East Asia were captured.

SILK SUBSTITUTE
Nylon, an artificial fibre that could be spun and woven, was first mass-produced in the 1940s. It was mainly used in stockings and underwear.

FLEXIBLE HARDNESS
These snooker balls are made of celluloid, a hard plastic. It is also flexible, and was used as a base for photographic film, and in men's shirt-collars.

Samples of isoprene

A BALATA BALL
Balata, a hard, rubber-like substance, is used for the outer casing of golf balls. Natural balata is now extremely rare, and has had to be replaced by a synthetic form.

Outer casing of golf ball is made of balata

IN THE FRAME
A tough, high-density form of polythene is sometimes used for spectacle frames. Polythene is most familiar as packaging.

MARBLED FOUNTAIN
Plastic cases made fountain-pens cheaper. Their "marbled" appearance was created by mixing different coloured plastics.

PLASTIC PLATTER
Bandelasta is a thermoset – it has been hardened by heating during its manufacture, and is consequently heat-resistant.

Slightly marbled bakelite

HOT STUFF
Bakelite is a good thermal and electrical insulator, and was used in items such as this 1930s hair-dryer.

PLASTIC PERSONALITY
The plastic Michelin Man advertises Michelin car tyres, which are made of vulcanized rubber.

SCREEN TEST
Chemists now work with molecules on computer screens. This is a molecule of enkephalin, a natural substance in the brain that affects the perception of pain. The atoms in the molecule are colour-coded, positioned, and in their correct proportions. Their positions can be modified, or new groups can be added. The computer has stored information about the forces between the atoms (pp. 60-61), so chemically impossible groupings are not permitted. The proposed molecule can be rigorously tested on screen, and precious laboratory research time can be saved.

Radioactivity

In 1880 THE ATOM WAS STILL THOUGHT TO BE impenetrable and unchanging. However, by 1900 this picture was seen to be incorrect. An important new discovery was radioactivity. This is the emission of invisible radiations by certain kinds of atom, happening spontaneously and unaffected by chemical reactions, temperature, or physical factors. The radiations are alpha, beta, or gamma (α, β, or γ). Ernest Rutherford (1871-1937) did most to clarify radioactivity. He found that α-particles were helium atoms, without electrons (pp. 48-49) and β-particles were fast electrons. When α- or β-particles were shot out from the atom, a different sort of atom was left. Such changes could cause γ-radiation, a type of electromagnetic radiation, to be emitted. Transmutation, long dreamed of by the alchemists as they tried to change one element into another, really was possible. It is now known that radiations in large doses, or in small exposures over long periods, can cause sickness and death. Nevertheless, radioactivity has many important uses. For example, metal objects can be "X-rayed" with γ-rays, medicines moving around the body can be tagged with radioactive "tracers", and archaeological finds can be dated by measuring their radioactivity.

BECQUEREL'S RAYS
While studying X-rays (radiation that could penetrate certain materials) Antoine Becquerel (1852-1908) stumbled on a new kind of invisible, penetrating radiation. In 1896 he found that crystals of a uranium compound could "fog" photographic film, even when the film was wrapped in black paper.

Cartoon of the Curies

A CURIOUS COUPLE
Marie Curie (1867-1934), assisted by her husband Pierre (1859-1906), found that the uranium ore pitchblende was considerably more radioactive than pure uranium. They realized that pitchblende must contain additional, more highly radioactive, substances. In 1902, after four years of laborious effort, they isolated tiny quantities of two new elements, polonium and radium. Like other early scientists working with radioactivity, the Curies knew little of its dangers, and Marie Curie died of leukemia. The high radiation levels with which she worked are evident from her glass flask – exposure to radiation turned it from clear to blue.

Marie Curie's glass flask

URANIUM ORE
Pitchblende is a brownish-black rock consisting mainly of uranium chemically combined with oxygen. It forms crystals called uraninite. Once considered useless, pitchblende is now the main source of uranium and radium.

FLASH GADGET
William Crookes (p. 48) invented the spinthariscope, for detecting α-particles. The α-particles struck a screen coated with zinc sulphide, creating a tiny flash seen through the eyepiece.

GEIGER'S GAUGE

Hans Geiger (1882-1945) gave this Geiger counter, a device for measuring radiation levels, to James Chadwick (pp. 52-53) in 1932. In this early model, low-pressure gas is contained in a copper cylinder, fitted with a handle. An electrical voltage is applied between this casing and a thin wire running along its centre. When an α- or β-particle enters the counter through a window at one end, it generates a brief burst of electric current between the case and the wire, which is detected on the counter.

Copper casing

Mica window

Thin wire runs the length of the counter

Insulated handle

Screw terminal

Connector

RADIATION OF THE ROCKS

A low level of "background" radioactivity is present in everything, even in our bodies. Levels are higher in regions of granite rock, for granite contains uranium. Granite emits radon gas, which can accumulate in homes and threaten health.

RADIOACTIVE SOLUTION

This liquid, uranyl nitrate, was prepared in 1905 by Frederick Soddy as part of his research into the transmutation of elements. It contains uranium and radium, and is highly radioactive. Its bright yellow colour is typical of uranium compounds.

Engraving on the flask reveals that the liquid contains 255g of purified uranium, and $16 \times 10^{-12}g$ of radium

PHYSICISTS AT WORK

Ernest Rutherford (right) and Hans Geiger in their laboratory at Manchester University, about 1908, with apparatus for detecting α-particles. Geiger and Rutherford realized that α-particles were helium atoms without electrons.

Part of image on shroud

CARBON-DATING THE TURIN SHROUD

The body of the crucified Christ was reputedly wrapped in a shroud, and was said to have created a life-size image still visible on the cloth. Analysis of a radioactive form of carbon taken from tiny samples of the shroud, now kept at Turin, showed that in fact the cloth was from medieval times.

Container for shroud sample

Archbishop of Turin's seal

Inside the atom

THE FIRST CLUE TO THE STRUCTURE of the atom came from experiments by J. J. Thomson (1856-1940) in 1897. He discovered particles that were smaller than atoms in cathode rays. These rays were seen passing between high voltage terminals in a glass tube filled with low pressure gas. The particles, called corpuscles by Thomson, and later known as electrons, had a negative electric charge and were about 2,000 times lighter than a hydrogen atom. They were exactly the same whatever gas was used in the tube, and whatever metal the terminals were made of. This strongly suggested that electrons were present in all matter. Atoms must also contain positive electric charge to balance the negative charge of electrons. Ernest Rutherford (pp. 46-47) probed atoms with particles produced in his experiments with radioactivity, and found that the positive charge was concentrated in a tiny nucleus. He reached the conclusion that the atom resembled a tiny solar system, where the "planets" were the electrons and the "Sun" was the nucleus.

MYSTERY RAYS
William Crookes (1832-1919) devised a glass tube that could contain a vacuum. It was used for the study of cathode rays (electrons emitted by a cathode – a negative terminal – when heated). He placed small obstacles in the rays, which cast "shadows", showing that their direction of travel was from the cathode to the positive terminal (the anode). They could make a small wheel turn in the tube, and Crookes concluded that the rays consisted of charged particles. The tube later became known as the Crookes tube.

High voltage between the metal plates creates an electric field, which bends the paths of charged particles

Paper scale for measuring deflection of electron beam

Low-pressure gas

Particles make glowing spot on glass

ATOM ATTACK
In 1911 Ernest Rutherford studied the effects of bombarding pieces of gold or platinum foil with alpha (α) particles – positively charged particles given out by radioactive materials (pp. 46-47). Most α–particles passed through the foil, but about one in 8,000 was deflected by more than 90°. Rutherford explained that this was due to the nucleus – a dense centre of positive charge within the atom.

Positively charged nucleus

α–particle deflected by more than 90°

α–particle scarcely deviated

α–particle track strongly bent

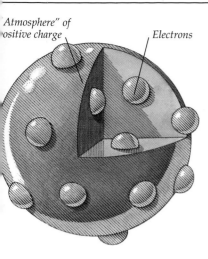

"Atmosphere" of positive charge

Electrons

THOMSON'S DISCOVERIES

J. J. Thomson intended to be a railway engineer, but instead became a brilliant physicist. He studied cathode rays with great success because he managed to achieve very low gas pressures in his modified Crookes tube. Thomson's discovery of the electron – the fundamental unit of electric current, present in all matter – revolutionized the theories of electricity and atoms. He also confirmed the existence of isotopes (pp. 52-53) – elements that each have several types of atom, chemically identical, but differing in weight.

THE PLUM PUDDING ATOM

In his plum pudding theory, J. J. Thomson suggested that every atom consists of a number of electrons, and an amount of positive charge to balance their negative charges. He thought the positive charge formed an "atmosphere" through which the electrons moved, like plums in a plum pudding.

Electrons pass through slits in anodes (positive terminal)

Heated cathode (negative terminal) produces electrons

WEIGHING THE ELECTRON

This is Thomson's original apparatus for studying cathode rays. It contained low pressure gas, through which cathode rays passed. The paths of the rays were bent by an electric field and Thomson measured the amount of bending. The electric field was switched off, a magnetic field was switched on, and again the bending was measured. Thomson calculated that if the particles had the same charge as the hydrogen ion (an "incomplete" atom) found in electrolysis (pp. 50-51) they must be about 2,000 times lighter.

Coils for magnetic field, which bends the charged particles

Negatively charged electron

RUTHERFORD'S DISCOVERIES

During his experiments with radioactivity, Ernest Rutherford discovered the transmutation (p. 46) of one element into another. He also studied the half life of an element – the time taken for half of a sample of a radioactive element to decay, or change into another element. He published his discoveries in 1904, in his book *Radioactivity*.

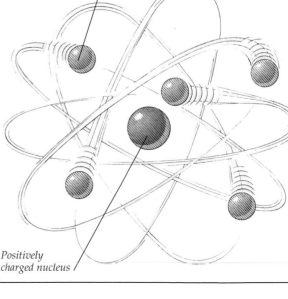

THE NUCLEAR ATOM

After Rutherford's explanation of the scattering of α–particles, the structure of the atom became clearer. Negatively charged electrons were thought to move around a positively charged, dense nucleus, rather like planets move around the Sun. However, there were problems with this "solar-system" model. According to the laws of physics at that time, such an atom should have collapsed instantly in a burst of electromagnetic radiation. (It is now known that the atom does not collapse because electrons are only "allowed" certain energies, pp. 50-51.)

Positively charged nucleus

Electrons, shells, and bonds

IN THE EARLY 1900S the structure of the atom became clearer, but the laws of physics at the time could not explain why electrons did not quickly spiral in towards the nucleus. Niels Bohr (1885-1962), a student of Rutherford, helped to solve the mystery by suggesting that electrons were only "allowed" certain energies. He found that electrons with the lowest allowed energy orbit closest to the nucleus, and electrons with the highest allowed energy orbit furthest away. It was soon discovered that there is a limit to the number of electrons with each energy. Electrons in an atom behave as if they are stacked, lowest energies first, in "shells" around the nucleus. It is the electrons in an atom's outermost shell that decide an atom's chemical properties. Atoms with outer shells "filled" with electrons are less reactive than those with only one electron in their outer shell. The outermost electrons join, or bond, with other atoms to form molecules. This new picture of the atom explained the reactions of atoms in processes such as electrolysis.

THE EXPENSE OF EROS
In 1884 this aluminium statue of Eros was highly expensive, but now aluminium can be cheaply produced by electrolysis.

LIBERATING LAWS
Michael Faraday (1791-1867) discovered the laws of electrolysis in 1833. He found that elements are liberated by a certain amount of electricity, or by twice or three times this amount. This depends on the number of outer electrons.

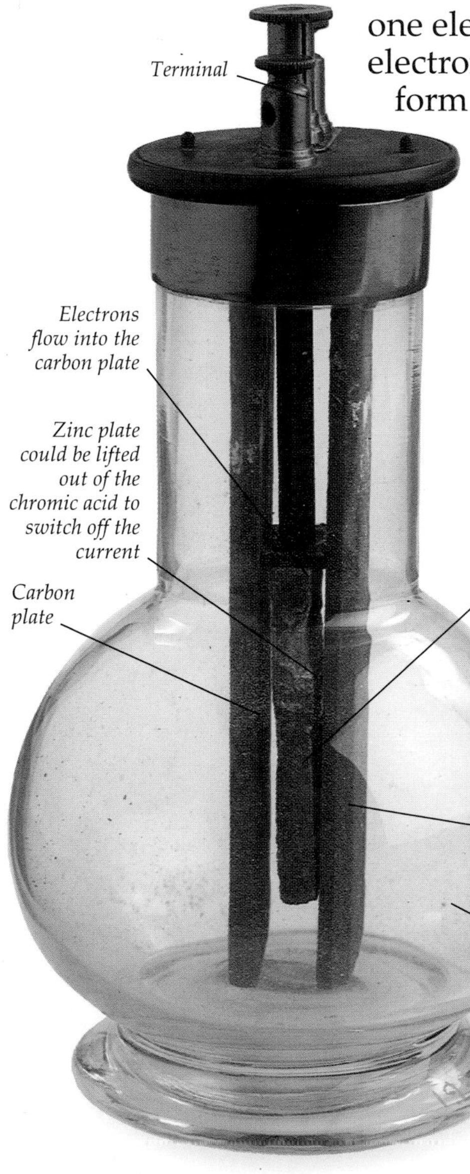

Terminal

Electrons flow into the carbon plate

Zinc plate could be lifted out of the chromic acid to switch off the current

Carbon plate

Electrons flow out of the zinc plate and round the electrical circuit connected to the terminals

Carbon plate

Glass jar was filled with chromic acid

ELECTRON STREAMS
Electric cells like this "bichromate cell" could be joined together to make a battery. The plates of carbon and zinc react with the chromic acid in the glass jar, transferring a stream of electrons from the zinc to the carbon.

ELECTRIFYING DISCOVERIES
In the 19th century many new elements were discovered by passing electric current through solutions or molten materials. The samples of metals shown here were prepared by electrolysis, and the electric current came from a battery. Electrolysis can separate compounds into elements by supplying electrons to, or removing them from, the outer shells of atoms.

Potassium.

Calcium.

Strontium.

Pure Iron

BREAKING DOWN WATER
This equipment was used by Michael Faraday (left) to study the decomposition of water by electricity. Hydrogen came off at one electrode and oxygen at the other. The amounts of these gases were measured, as was the amount of electricity required to release them from water. Faraday worked out the laws of electrolysis in this way. The basic amount of electricity used in electrolysis is called the Faraday constant.

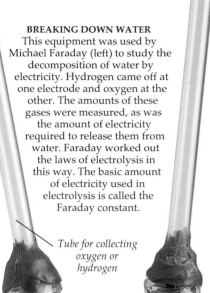

Tube for collecting oxygen or hydrogen

Terminals were connected to a battery

Glass globe was filled with water and a little acid

Platinum electrode

LIGHT ON THE MATTER
Niels Bohr explained the connection between matter and light in 1913. He suggested that when electrons move from one energy level to another, they give out or absorb "packets" of radiation in the form of light. These packets are called photons, or quanta. The shorter the radiation's wavelength, the higher the photon's energy.

Eight electrons in second level or "shell"

Lone electron

Two electrons in first level or "shell"

Nucleus

BOHR'S ATOM
In Bohr's theory of the atom, electrons that are further out from the nucleus have higher energy, and an electron can jump to a higher level by absorbing energy. This can happen at high temperatures, or when photons with enough energy hit the atom. If there is a gap in a lower level, an electron can fall down to that level, giving out energy in the form of radiation.

IONIC BONDS
Sodium and chloride ions are held together by their opposing electric charges. The sodium atom "wants" to lose its outer electron because the atom is unstable, and the incomplete atom (ion) is left with a positive electric charge. A chlorine atom "wants" to gain an electron to fill its outer shell, and gains an extra negative charge.

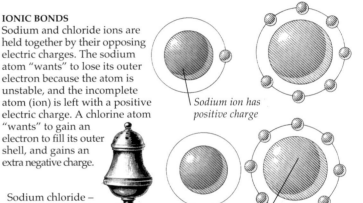

Sodium ion has positive charge

Chloride ion has negative charge

Sodium chloride – common salt

Mask to avoid poisoning from chlorine gas

COVALENT BONDS
Atoms can share electrons in their outer shells to produce filled shells, forming a "covalent" bond. Chlorine atoms, with seven electrons in the outer shell, can pair off, each pair sharing two electrons. Each atom effectively has eight electrons in the outer shell. Like many other gases, chlorine normally exists in the form of two-atom molecules. The bond is easily broken, making chlorine reactive and dangerous.

Chlorine atoms can pair off and share two electrons

"Pool" of shared electrons

Aluminium aircraft

METALLIC BONDS
The atoms of metals share their outer electrons. They are contributed to a "pool", and wander freely from atom to atom. The electrons' ease of movement is why metals are good conductors of heat and electricity.

CLOUDS OF MYSTERY
Bohr's sharply defined electron orbits have been superseded by fuzzy electron "clouds" (right), which can be seen with an electron microscope. It is now known that electrons behave as waves, as well as like particles. An electron is most likely to be found where the electron "cloud" is dense. But there is always a definite, if small, chance of finding it closer to, or farther from, the nucleus.

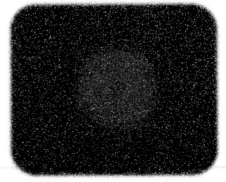

Architecture of the nucleus

By the early 20th century, it was known that the atom had a positively charged nucleus. Ernest Rutherford (pp. 46-47) suggested that the nucleus contained positively charged particles called "protons" (Greek for "first things"). He demonstrated their existence in 1919 by knocking them out of nitrogen nuclei using alpha (α) particles. James Chadwick (1891-1974) discovered another particle in the nucleus in 1932 – the neutron, an uncharged particle of about the same mass as the proton. All nuclei comprise protons and neutrons. The number of protons determines the number of electrons circling the nucleus, and therefore the chemical properties of the atom (pp. 50-51). All elements have different isotopes – atoms with the same number of protons but different numbers of neutrons.

PARTICLE DISCOVERER
James Chadwick, a student of Rutherford, discovered neutrons by exposing the metal beryllium to α-particles. He observed a new kind of particle ejected from its nucleus, the neutron. Later he studied deuterium (also known as "heavy hydrogen"). This isotope of hydrogen was discovered in 1932 and is used in nuclear reactors.

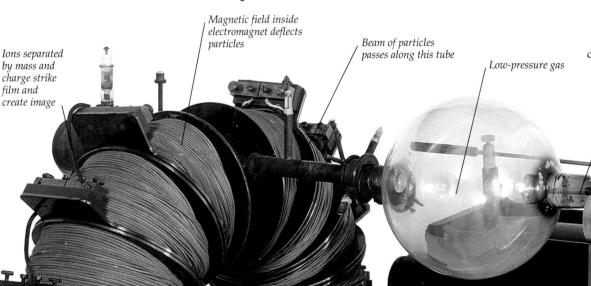

Magnetic field inside electromagnet deflects particles

Beam of particles passes along this tube

Low-pressure gas

Ions separated by mass and charge strike film and create image

Ions of particular kind are produced here

Anode (positive terminal)

SEPARATING ISOTOPES
This was the first mass spectrograph, designed by F. W. Aston (1877-1945). It could separate isotopes – chemically identical atoms with different masses. The globe contained a compound of the material to be tested, either as part of the anode (positive terminal) or as low pressure gas. An electric current knocked electrons from the material's atoms, leaving positively charged ions that passed through a collecting slit. The beam of charged particles was bent by an electric field and then by a magnetic one. It was spread into separate bands on a photographic film, according to the ions' charge and mass.

Electrons, protons, and neutrons

Rutherford believed the nucleus was made up of protons and a smaller number of electrons. He thought that each electron was closely paired with a proton to make a "doublet" that was neutral (had zero electric charge). In 1932 James Chadwick produced a type of radiation that did not bend in an electric field, but was far more penetrating than gamma rays. This radiation consisted of uncharged particles, known as "neutrons", that were about as massive as hydrogen atoms. Chadwick realized that these neutrons might be particles themselves, and not a proton and electron combined. This view is now accepted. However, a free neutron has a 50 per cent chance of "decaying" into a proton and electron in 15 minutes. If a proton and electron collide they will produce a neutron.

NEUTRON DETECTOR
Inside this intriguing apparatus built by Chadwick, α-particles from a radioactive source struck a beryllium target. The neutrons given off could only be detected when they knocked protons from a piece of paraffin wax. The protons were detected with a Geiger counter (pp. 46-47).

Tube was fixed to an air pump to take air out of the chamber

Chamber contained a radioactive source

A SCIENTIST'S TOOLS
This cigarette carton was Chadwick's "toolbox". He used the pieces of paraffin wax to observe neutrons. The silver and aluminium foils, of different thicknesses, were used as barriers to determine the penetrating power of the radiation.

RHODIAN
CIGARETTES

Power supply

One of six orbiting electrons – negative charge

CARBON-12
The chemical properties of carbon are determined by its six negatively charged electrons. These six electrons balance the six positively charged protons of the nucleus. In a carbon-12 atom, the nucleus also has six neutrons, of about the same mass as the protons, giving the atom a "mass number" of 12.

One of six protons – positive charge

One of six neutrons – zero charge

One of six orbiting electrons – negative charge

CARBON-14
The isotope carbon-14 is chemically identical to ordinary carbon. It has six protons and six electrons. But carbon-14 has two extra neutrons, giving it a mass number of 14. This isotope is radioactive, 50 per cent of it decaying every 5,730 years. Environmental levels are roughly constant, since new carbon-14 atoms are constantly being formed by cosmic rays smashing into ordinary carbon atoms.

One of six protons – positive charge

One of eight neutrons – zero charge

Splitting the atom

AFTER THE DISCOVERY of the nucleus in 1911, it was found that bombarding certain atoms with particles from radioactive materials could disintegrate their nuclei, releasing energy. The heaviest nuclei, those of uranium, can be split by neutrons in this way. Otto Hahn (1879-1968) and Lise Meitner (1878-1968) discovered that the uranium nucleus splits in half, or "fissions", and also gives out further neutrons. These neutrons can go on to cause further fissions. In 1942 a team led by Enrico Fermi (1901-1954) achieved this "chain reaction" in the world's first nuclear reactor. Three years later the chain reaction was used in the nuclear bombs that destroyed the Japanese cities of Hiroshima and Nagasaki.

Neutrons produced by fission

Typical products of fission are barium and krypton

Nucleus fissions into two smaller nuclei

CHAIN REACTION
The source of energy in a nuclear reactor or a nuclear explosion is a chain reaction. A nucleus of uranium or plutonium splits (fissions), giving out neutrons that split further nuclei. Immense heat is created by the energy of the splitting fragments and by radiation. In a reactor this heat is used in a controlled way to generate electricity. In an explosion it is released violently.

2 NUCLEAR SPLIT
When a neutron hits another uranium nucleus, the nucleus fissions into two smaller nuclei of approximately half the size. Several neutrons are also given out, together with high-energy radiation. The neutrons can go on to cause further fissions in a chain reaction. Neutrons can be slowed down by graphite or heavy water mixed with the uranium.

Uranium nucleus (U-235)

Neutron about to hit uranium nucleus

1 WANDERING NEUTRONS
Neutrons can be released by bombarding atoms with radiation. Neutrons are also occasionally given out by decaying uranium nuclei, but these neutrons rarely react with uranium nuclei to build a chain reaction. Many nuclear reactors use uranium-235, a very reactive but uncommon isotope.

Unstable uranium
The main isotope of uranium is uranium-238 (symbol U-238). It has 238 particles in its nucleus – 92 protons and 146 neutrons. The neutrons prevent the protons from blowing the nucleus apart because of their mutual repulsion. Even so, an unstable U-238 nucleus breaks down from time to time, giving out an alpha (α) particle and turning into a thorium nucleus. The thorium nucleus in turn breaks down, and so do its products, in a chain of decays that ends when a lead nucleus is formed. Other uranium isotopes go through a similar chain of radioactive decays, ending in a different isotope of lead. This is why uranium-bearing rocks can be detected by their radioactivity. Uranium can also break down by fission, and this process can build up into a chain reaction. For a chain reaction to occur, there must be special conditions, and a sufficient quantity of relatively pure uranium must be used.

FAMILY FISSION
In 1917 Lise Meitner and Otto Hahn discovered a new element, protactinium, found in uranium ores. In 1939 Meitner and her nephew Otto Frisch (1904-1979) announced the fission of uranium.

PUZZLING PRODUCT
Otto Hahn studied the disintegration of uranium nuclei by neutrons. Among the by-products of this disintegration were barium nuclei, which are about half the weight of uranium nuclei.

HEAVY WATER
Neutrons in a nuclear reactor's chain reaction can be controlled by a moderator such as heavy water. It is 11 per cent heavier than an equal volume of ordinary water.

NORSK HYD KVÆLSTOF

DEUTERI Gr d.²⁰

54

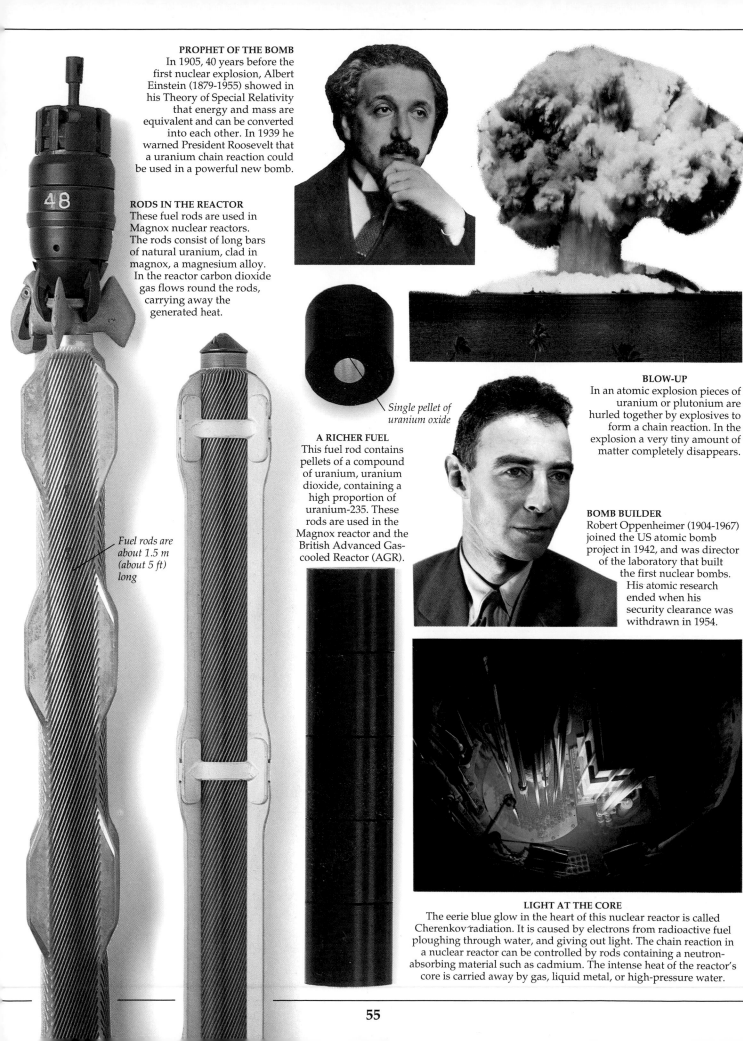

PROPHET OF THE BOMB
In 1905, 40 years before the first nuclear explosion, Albert Einstein (1879-1955) showed in his Theory of Special Relativity that energy and mass are equivalent and can be converted into each other. In 1939 he warned President Roosevelt that a uranium chain reaction could be used in a powerful new bomb.

RODS IN THE REACTOR
These fuel rods are used in Magnox nuclear reactors. The rods consist of long bars of natural uranium, clad in magnox, a magnesium alloy. In the reactor carbon dioxide gas flows round the rods, carrying away the generated heat.

Single pellet of uranium oxide

A RICHER FUEL
This fuel rod contains pellets of a compound of uranium, uranium dioxide, containing a high proportion of uranium-235. These rods are used in the Magnox reactor and the British Advanced Gas-cooled Reactor (AGR).

Fuel rods are about 1.5 m (about 5 ft) long

BLOW-UP
In an atomic explosion pieces of uranium or plutonium are hurled together by explosives to form a chain reaction. In the explosion a very tiny amount of matter completely disappears.

BOMB BUILDER
Robert Oppenheimer (1904-1967) joined the US atomic bomb project in 1942, and was director of the laboratory that built the first nuclear bombs. His atomic research ended when his security clearance was withdrawn in 1954.

LIGHT AT THE CORE
The eerie blue glow in the heart of this nuclear reactor is called Cherenkov radiation. It is caused by electrons from radioactive fuel ploughing through water, and giving out light. The chain reaction in a nuclear reactor can be controlled by rods containing a neutron-absorbing material such as cadmium. The intense heat of the reactor's core is carried away by gas, liquid metal, or high-pressure water.

Hot matter

ATOMS ARE LAID BARE as they are subjected to high temperatures. The spectroscope reveals their secrets by analysing the light they give out. In a spectroscope the light falls on a diffraction grating – a flat surface with thousands of lines – or a prism. Light passes through or is reflected, and is broken up into different colours. Sunlight consists of the whole colour spectrum and beyond. Gases at the Sun's surface produce sunlight, at temperatures of about 5,500° C (about 10,000° F). Here, the atoms' outer electrons are knocked to higher orbits and give out light as they fall back (pp. 50-51). Inside the Sun and other high-temperature stars, inner electrons are knocked to higher orbits. As they fall back, they give out ultraviolet and X-rays. At the centres of the Sun and stars, at temperatures of around 15 million° C (about 27 million° F), nuclei are stripped bare and welded together, producing heavier nuclei.

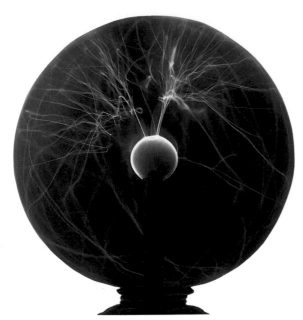

INSIDE THE PLASMA BALL
A powerful voltage at the centre of this glass globe tears electrons from the atoms of the low pressure gases inside. Avalanches of electrons build up, forming bright squiggly lines of hot gas. The mixture of electrons and charged atoms in these lines is called a plasma.

A LIGHT SIGNATURE
The spectrometer "reads the signature" of materials by analysing their light. The light first passes through a narrow slit into a small telescope, which focuses the light into a narrow parallel beam. The beam passes through a glass prism, and is spread out, with each wavelength (colour) of light going in a slightly different direction. Through a viewing telescope a rainbow-coloured "spectrum" can be seen. This may appear as a host of bright lines, or a continuous band of colour, crossed by dark lines, where wavelengths have been absorbed.

Photographic plate

Plate-holder

Camera

THE VANISHING SUN
The Sun is kept at a high temperature by the flood of energy from the nuclear reactions at its heart. Every second, four million tonnes of the Sun's matter vanish, converted into energy that escapes from the surface as radiation.

ABSORBING EXPERIMENT
This 19th-century spectroscopic experiment shows light from a gas flame passing through a liquid containing dissolved materials. The spectrum that is produced reveals the identity of the dissolved materials in the liquid.

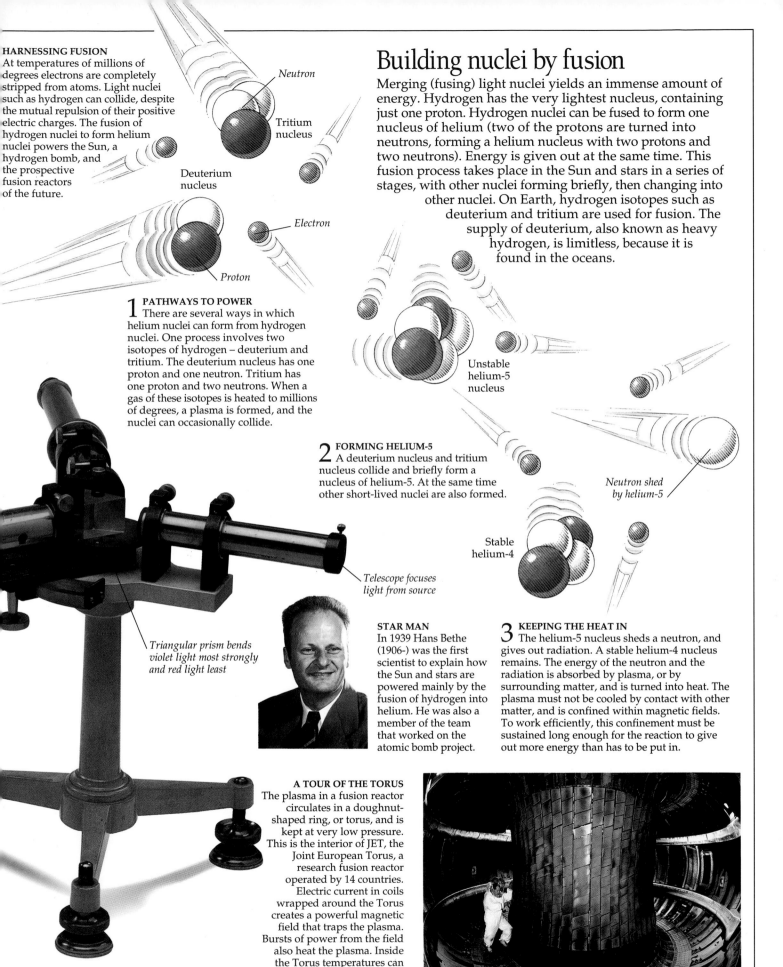

HARNESSING FUSION
At temperatures of millions of degrees electrons are completely stripped from atoms. Light nuclei such as hydrogen can collide, despite the mutual repulsion of their positive electric charges. The fusion of hydrogen nuclei to form helium nuclei powers the Sun, a hydrogen bomb, and the prospective fusion reactors of the future.

Neutron

Tritium nucleus

Deuterium nucleus

Electron

Proton

Building nuclei by fusion

Merging (fusing) light nuclei yields an immense amount of energy. Hydrogen has the very lightest nucleus, containing just one proton. Hydrogen nuclei can be fused to form one nucleus of helium (two of the protons are turned into neutrons, forming a helium nucleus with two protons and two neutrons). Energy is given out at the same time. This fusion process takes place in the Sun and stars in a series of stages, with other nuclei forming briefly, then changing into other nuclei. On Earth, hydrogen isotopes such as deuterium and tritium are used for fusion. The supply of deuterium, also known as heavy hydrogen, is limitless, because it is found in the oceans.

1 PATHWAYS TO POWER
There are several ways in which helium nuclei can form from hydrogen nuclei. One process involves two isotopes of hydrogen – deuterium and tritium. The deuterium nucleus has one proton and one neutron. Tritium has one proton and two neutrons. When a gas of these isotopes is heated to millions of degrees, a plasma is formed, and the nuclei can occasionally collide.

Unstable helium-5 nucleus

2 FORMING HELIUM-5
A deuterium nucleus and tritium nucleus collide and briefly form a nucleus of helium-5. At the same time other short-lived nuclei are also formed.

Neutron shed by helium-5

Stable helium-4

Telescope focuses light from source

Triangular prism bends violet light most strongly and red light least

STAR MAN
In 1939 Hans Bethe (1906-) was the first scientist to explain how the Sun and stars are powered mainly by the fusion of hydrogen into helium. He was also a member of the team that worked on the atomic bomb project.

3 KEEPING THE HEAT IN
The helium-5 nucleus sheds a neutron, and gives out radiation. A stable helium-4 nucleus remains. The energy of the neutron and the radiation is absorbed by plasma, or by surrounding matter, and is turned into heat. The plasma must not be cooled by contact with other matter, and is confined within magnetic fields. To work efficiently, this confinement must be sustained long enough for the reaction to give out more energy than has to be put in.

A TOUR OF THE TORUS
The plasma in a fusion reactor circulates in a doughnut-shaped ring, or torus, and is kept at very low pressure. This is the interior of JET, the Joint European Torus, a research fusion reactor operated by 14 countries. Electric current in coils wrapped around the Torus creates a powerful magnetic field that traps the plasma. Bursts of power from the field also heat the plasma. Inside the Torus temperatures can reach as high as 300 million° C (about 550 million° F).

Subatomic particles

In the early 1930s the atom seemed to be made of three kinds of particle – the proton, neutron, and electron. But soon more particles were found. The existence of the neutrino – a ghostlike particle that "carries away" energy when a neutron decays (p. 53) – was suspected. Then the muon, rather like a heavy electron, and the pion, which binds protons and neutrons together in the nucleus, were both discovered in cosmic rays. Accelerators were built to smash particles into nuclei at high speed, creating new particles. Today hundreds of particles are known. They seem to fall into two main classes, hadrons and leptons. Hadrons include the proton and neutron, and are made of pairs or triplets of quarks, which are never seen singly. Leptons, the other class, include electrons and neutrinos.

(p. 53)

MAKING TRACKS
Trails of water droplets in a cloud chamber mark the paths of electrons and positrons (similar to an electron but with a positive charge). Because of their opposite electric charges, they curve in different directions in the chamber's magnetic field. An electron produced near the bottom of the picture circles 36 times before losing its energy.

AMAZING REVELATIONS
The cloud chamber, invented by Charles Wilson (1869-1959) in 1911, was the first detector to reveal subatomic particles in flight. Particles from a radioactive source pass through a glass chamber, which contains air and water vapour. In the glass chamber, particles knock electrons out of the atoms in the air, leaving positively charged ions (incomplete atoms). The pressure in the chamber is suddenly reduced, and water vapour condenses on the ions, forming trails of small drops.

Chamber contains water vapour

The piston at the bottom of the chamber moves to form water vapour, which condenses on the tracks of particles as they pass through

To lower the piston the flask is emptied of air – the connection between the flask and the space below the piston is opened and the piston is suddenly sucked downwards

Glass negatives of cloud chamber photographs

PARTICLE PICTURES
Photographic plates of cloud chamber tracks often show particles being created and destroyed. Measuring the tracks can reveal the particles' electric charge, mass, and speed.

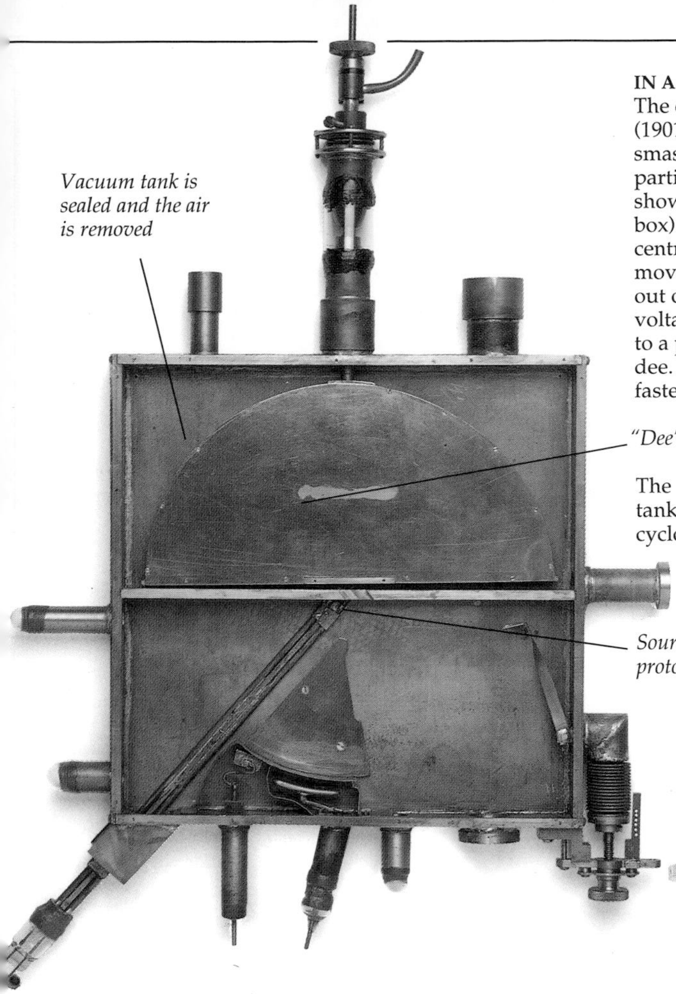

IN A WHIRL

The cyclotron, invented by Ernest Lawrence (1901-1958) in 1930, accelerated particles and smashed them into atomic nuclei to form new particles. The vacuum tank of the cyclotron shown here housed a metal "dee" (a D-shaped box). Charged particles entered the dee at the centre of the apparatus, and a magnetic field moved them in a small circle, half in and half out of the dee. A rapidly varying electric voltage was applied to the dee, giving a "kick" to a particle whenever it left or re-entered the dee. The particle spiralled outwards, travelling faster and faster, until it left the cyclotron.

Vacuum tank is sealed and the air is removed

"Dee"

The vacuum tank of the cyclotron

Source of protons

JESTER OF PHYSICS

Richard Feynman (1918-1988) shared a Nobel Prize in 1965 for his work on the forces between particles and electromagnetic radiation. He was considered a brilliant teacher, and was famous for his practical jokes.

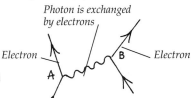

Photon is exchanged by electrons

Electron　　A　　B　　*Electron*

FEYNMAN DIAGRAM

This strange squiggle is a Feynman diagram. It illustrates that electromagnetic force between electrons occurs when they exchange a photon (the "carrier" of the electromagnetic force, pp. 60-61).

PARTICLE FROTH

This bubble chamber, made in 1956, contained liquid hydrogen at low temperature and high pressure. The pressure was suddenly released, and particles passed through the chamber. The liquid boiled on the trails of charged atoms left by the particles. The trails were photographed, and the chamber was quickly repressurized.

Bubble chamber contained liquid hydrogen

MAGIC CIRCLE

The underground Tevatron, in Illinois, USA, has two accelerators, one above the other. The upper one feeds particles to the more powerful lower one.

BEAUTY IN DECAY

In this false-colour bubble-chamber image, a high-speed proton (yellow, bottom) collides with a proton in a hydrogen atom and disappears, creating a shower of particles. An uncharged particle called a lambda leaves no track, but reveals itself by "decaying" into a proton and a pion (yellow and purple, centre).

Lower chamber

"Porthole" for viewing tracks

When in use, the bubble chamber sat in the lower chamber and was kept at a low temperature

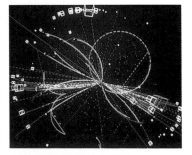

RECONSTRUCTING THE EVENT

Computers are now used to reconstruct subatomic events. This is a computer simulation of the decay of the Z^0 particle, one of the carriers of the weak nuclear force (pp. 60-61).

The four forces

NOT-SO-WEAK FORCE
The Sun is powered by the weak nuclear force, which is responsible for the conversion of hydrogen into helium at the Sun's core (p. 57). Under the less extreme conditions on Earth, the weak nuclear force is involved in radioactivity. This force does not extend beyond the atomic nucleus and could not be detected until scientists had learned how to probe inside the atom. The particles that carry the weak force, the W^+, W^-, and Z^0, were discovered in 1983 among the debris formed when subatomic particles collided in a giant accelerator.

ALL MATTER IS SUBJECT TO four forces – gravity, electromagnetism, and the weak and strong nuclear forces. Gravity holds people on the Earth and the planets in orbit around the Sun. Electrons are held in atoms by electromagnetism, a force that is enormously stronger than gravity. The weak nuclear force, a hundred thousand million times weaker than electromagnetism, is involved in radioactivity and nuclear fusion (pp. 56-57). The strong nuclear force, a hundred times stronger than electromagnetism, affects particles called quarks. Protons, neutrons, and other particles are made up of pairs or triplets of quarks. Electromagnetism is "carried" by particles called photons, the weak nuclear force by W and Z particles, and the strong nuclear force by particles called gluons. Gravity is probably carried by particles, too – these have been dubbed gravitons. Electricity and magnetism are "unified" because electricity in motion produces magnetic fields, and changing magnetic fields produces electrical voltages. Electromagnetism is in turn unified with the weak nuclear force, because at extremely high energies and temperatures they merge into one "electro-weak" force. The evidence for this comes from ideas about the first moments of the Big Bang (pp. 62-63) and from experiments in particle accelerators. Physicists are now working to develop a theory in which all four forces would be aspects of one superforce.

Earth

Moon

Sun

An orrery, a mechanical model of planetary motions

PULLING POWER
Gravity is the force in control of the entire solar system – it holds planets, asteroids, comets, and smaller bodies in orbit. The farthest known planet, Pluto, is firmly held by gravity even when it is over 7,000,000,000 km (about 4,000,000,000 miles) away from the Sun. Gravity extends far beyond this however – clusters of galaxies millions of light-years across are held together by their own gravity. Yet it is by far the weakest of the four forces. It dominates the universe because it is long-range, whereas the far stronger nuclear forces do not extend beyond the nucleus. Gravity is cumulative – it always attracts, never repels, so when matter accumulates into planet-sized or star-sized objects, a large gravitational force is developed. Electromagnetic forces are also long-range, but unlike gravity can both attract and repel, and generally cancel themselves out.

EVERYDAY INTERACTIONS

Many forces can be easily perceived, such as the way materials hold together and the friction between objects. These are both examples of the electromagnetic force. Gravity is the other force that is most obvious in human life. The electromagnetic force and gravity are shown in the following sequence.

1 GETTING THINGS GOING

A ball falls to the floor because the Earth's gravity pulls it, but the ball also pulls the Earth with exactly the same force. However, since the Earth has so much more mass, it does not move detectably, while the ball moves faster and faster. It is described as gaining energy of movement, or kinetic energy. Energy can be defined as the capacity to make things happen – for example to break things, to make them hotter, and to set them moving.

QUARK-FINDER

The strongest force of all, the strong nuclear force, is only felt by quarks. It binds them tightly together, and they have never yet been observed singly. During the 1980s evidence about the strong force carrying quarks, and the weak force, came from experiments in this giant accelerator, called the Super Proton Synchrotron.

EXCHANGING MESSAGES

When two particles are interacting through one of the four fundamental forces, they constantly exchange messenger particles. The messenger particles influence the other particles' movements rather as a tennis ball influences the tennis players' movements. The force can be a repulsion or an attraction.

3 GREAT POTENTIAL

The potential energy that was momentarily stored in the ball is converted into kinetic energy, and the ball shoots upwards. As the ball rises, it loses kinetic energy. If it should fall back to the ground later it will regain its speed. It is therefore described as converting kinetic into potential energy as it rises. In this case the potential energy is associated with height above the ground.

Gravity pulls the ball to the floor, but the ball pulls the Earth with exactly the same force

Potential energy is converted into kinetic energy and the ball bounces up

2 DOWN WITH A BUMP

When the ball hits the floor, the force of gravity is opposed by electromagnetism. The electrons in the outer layers of atoms in the ball and the floor repel each other. The upward push of electromagnetism overcomes the downward pull of gravity. The motion is abruptly stopped, but the ball's kinetic energy is converted into other forms. Some is dispersed through the material of the ball and floor as heat. Some is stored as potential energy (energy waiting to be released) in the ball. The electromagnetic forces between the atoms are distorted by the impact, and try to restore the ball to its normal shape. When they succeed the ball regains its kinetic energy.

CONSTANT QUANTITY

In any isolated system the total quantity of mass and energy is conserved. In a steam engine, for example, chemical energy of the fuel is converted into heat energy of the fire and of the steam. This heat energy is in turn converted into kinetic energy of the wheels driven by the engine. The total amount of mass and energy is always conserved whichever of the four fundamental forces are involved.

The birth and death of matter

THE TOTAL AMOUNT of mass and energy in the universe never changes. According to a widely held theory, billions of years ago the universe contained matter and energy of extraordinarily high density and temperature, which exploded in the Big Bang: As the gas expanded and cooled, quarks formed protons and neutrons, and some of these built helium nuclei. Eventually complete hydrogen and helium atoms formed. The gas condensed into galaxy-sized clouds, which broke up into stars. In the far future, the universe could collapse, be rejuvenated in a new Big Bang, and re-expand, but it is more likely that it will always expand. After the last star has faded, even protons may decay into much lighter particles, and the universe may end as a sea of electrons, neutrinos, and forms of radiation.

VIOLENT UNIVERSE
Early astronomers thought that the stars were tranquil and unchanging. It is now known, however, that they are born, lead violent lives, and die.

PUTTING THE CLOCK BACK
Time does not necessarily go forwards, or even at the same speed. If the universe were to collapse, it is possible that time could go backwards. Time slows down for high-speed objects – a cosmonaut in orbit for a year ages less (by a hundredth of a second) than people on Earth. Even time travel may be possible. In theory, two regions of the universe can be connected by a "wormhole", passing though other dimensions. An object entering one end of a wormhole could reappear from the other end, at an earlier time.

Time could go backwards in the extreme conditions found in the cosmos

SCATTERED SEED
The Crab Nebula is a mass of gas from a supernova – the explosion of a giant star, seen by Chinese astronomers in 1054. The gas is rich in elements made in the star's core. These will be scattered through space and some will be incorporated into new planets as they are born. All the elements in our bodies were made in some ancient supernova.

THE GRAVITY OF DARK MATTER

Galaxies are huge collections of stars, gas, and dust. Light, travelling at 300,000 km (about 200,000 miles) per second, can take as long as 100,000 years to cross a galaxy. Galaxies are grouped into clusters, which hurtle apart. There may be undetected dark matter in the vast spaces between galaxies. The gravity of dark matter may be sufficient to slow the galaxies' expansion and turn it into a collapse.

A BRIEF HISTORY OF BLACK HOLES

Stephen Hawking (1942-) is renowned for his work on explaining the birth of the universe and also for his theories about black holes. When matter becomes extremely dense, as in the core of an exploded star, its gravity becomes so powerful that both matter and radiation, including light, are trapped inside. The result is a black hole. Hawking showed that a black hole gives off radiation very slowly. His most popular work is the 1988 book *A Brief History of Time*.

AN INTELLIGENT UNIVERSE?

Fred Hoyle (1915-) is associated with the "steady state" theory, which holds there was no Big Bang, and matter is always being created throughout space. He claims that some physical laws have been designed by a superior intelligence to produce conditions that make the development of carbon-based life possible.

BACK TO THE BEGINNING

So far, the strongest evidence available to scientists for the Big Bang theory is cosmic background radiation. This is microwave radiation that can be detected by large radio telescopes like the one shown here. The radiation always comes with the same strength from all directions in the sky, and is believed by some to have travelled through space since the universe was 100,000 years old. Until this time the universe is thought to have been made of hot expanding plasma (pp. 56-57). The plasma then cooled sufficiently to allow electrons and nuclei to join up and form the first complete atoms.

Index

Acknowledgments

Dorling Kindersley would like to thank:
John Becklake, Chris Berridge, Tim Boon, Roger Bridgman, Neil Brown, Robert Bud, Sue Cackett, Ann Carter, Ann de Caires, Helen Dowling, Stewart Emmens, Robert Excell, Stephen Foulger, Graeme Fyffe, Derek Hudson, Kevin Johnson, Sarah Leonard, Steve Long, Stephanie Millard, Peter Morris, Keith Parker, David Ray and his staff, Anthony Richards, Derek Robinson, Victoria Smith, Peter Stephens, Laura Taylor, Peter Tomlinson, Denys Vaughan, Nicole Weisz, and Anthony Wilson for help with the provision of objects for photography; Jane Bull for design guidance; Deborah Rhodes for page makeup; Debra Clapson for editorial assistance; Marianna Papachrysanthou for design assistance; Neil Ardley for synopsis development; Stephen Pollock-Hill at Nazeing Glassworks for his expertise; Jane Burton, Jane Dickins, Paul Hammond, and Fiona Spence of De Beers for providing props.

Picture research Deborah Pownall and Catherine O'Rourke
Illustrations John Woodcock
Index Jane Parker

Picture credits

t=top b=bottom c=center l=left r=right

Bildarchiv Preussischer Kulturbesitz 54bl. Bridgeman Art Library /Royal Institution 32tr; /Chateau de Versailles /Giraudon 42tr. British Film Institute 14cr. British Library, London 16tl; 26cl. Brown Brothers 51tc. Cavendish Laboratory, Cambridge 49tr. Camera Press 43tr; 52tr; 63cb; 63cr. E. T. Archive 11c. Mary Evans Picture Library 10tl; 10br; 16bl; 18tr; 22c; 26clb; 39tc; 44tr; 44c. W. H. Freeman and Co., from Powers of Ten by Philip Morrison and the Office of Charles and Ray Eames. Copyright (c) 1982 by Scientific American Library 51br. Robert Harding Picture Library 17tl; 27bc. Michael Holford 41c. Hulton Picture Co. cover back cl; 8bl; 33cr; 38tr; 40c; 50tr; 49bl; 54br. Jet Joint Undertaking 57br. Mansell Collection 8tr; 14tl; 47bl; 48tl; 55tc. National Portrait Gallery, London 26bl. Oxford Scientific Films /Manfred Kage 38tl. Ann Ronan Picture Library cover front cr; 13tl; 13tr; 19cr; 23tr; 42c; 46tl; 46cl. Science Museum Photo Library 12bl; 15cl; 16crb; 21tl; 21tc; 21cr; 27tc; 28cr; 30cl; 34cl; 34b; 35tc; 35tr; 35br; 37tl; 51c; 56br. Science Photo Library /Brookhaven National Laboratory cover front c/James Bell 15br; /Keith Kent 19tl; /Claude Nuridsany and Maria Perendu 19bc; /Peter Manzell 23tc; /Andrew McClenaghan 39tr; /Clive Freeman 41bc; /Jeremy Burgess 43br; / J. C. Revy 45br; 47c; 58tr; 59cl; 59bl; 59tr; 59br; 62bl; /Peter Menzel 63br; /US National Archives 55tr; /Los Alamos National Laboratory 55cr; /Sandia National Laboratories 55br; /Los Alamos National Laboratory 57cl; /Barney Magrath 60tl; /Heini Schneebeli 61tr. Roger Viollet 28bl. Ullstein Bilderdienst 39c. Zefa 14tr; 29cl; 50tl; 63tr.

With the exception of the items listed above and the following: 3tl; 6-7, 11bl; 12-13, 14cr, 15tl, 15bl, 18-19, 24cl, 24bl, 24cr, 25, 28-29, 37tr, 44bc, 44t, 44tr, 45, 60tr, 61, 62-63, all the photographs in this book are of objects in the collections of the Science Museum, London.